COVER PHOTOGRAPH

Space Needle
Originally constructed for the 1962 World's Fair, the Space Needle in Seattle, WA has become a symbol representing the city as well as the Northwest region of the country. The tower is over 600 feet tall with an accessible observation deck and restaurant over 500 feet above the ground.
©iStockphoto.com/antony spencer

Copyright ©2014 by SM^c Curriculum, LLC. All rights reserved. Printed in the United States. This publication is protected by copyright. No part of this publication should be reproduced or transmitted in any form or by any means without prior written consent of the publisher. This includes, but is not limited to, electronic reproduction, storage in a retrieval system, photocopying, recording or broadcasting for distance learning. For information regarding permission(s), write to: Permissions Department.

ISBN: 978-1-938801-75-4

2 3 4 5 6 7 8 9 10

ABOUT THE AUTHORS

From left to right: Beth Armstrong, Matt McCaw, Shannon McCaw, Scott Valway, Michelle Terry, Sarah Schuhl

SERIES AUTHOR

Shannon McCaw has been a classroom teacher in the Newberg and Parkrose School Districts in Oregon. She has been trained in Professional Learning Communities, Differentiated Instruction and Critical Friends. Shannon currently works as a consultant with math teachers from over 100 districts around Oregon. Shannon's areas of expertise include the Common Core State Standards, curriculum alignment, assessment best practices and instructional strategies. She has a degree in Mathematics from George Fox University and a Masters of Arts in Secondary Math Education from Colorado College.

CONTRIBUTING AUTHORS & EDITORS

Beth Armstrong has been a classroom teacher in the Beaverton School District in Oregon. She has received training in Talented and Gifted Instruction. She has a Masters in Curriculum and Instruction from Washington State University.

Matt McCaw has been a classroom teacher, math/science TOSA and special education case-manager in several Oregon school districts. Matt has most recently worked as a curriculum developer and math coach for grades 6-8. He is trained in Differentiated Instruction, Professional Learning Communities, Critical Friends Groups and Understanding Poverty. Matt has a Masters of Special Education from Western Oregon University.

Sarah Schuhl has been a classroom teacher, secondary math instructional coach and district-wide K-12 math instructional specialist, most recently in the Centennial School District in Oregon. Sarah currently works as a Solution Tree associate and an educational consultant supporting and challenging teachers in the areas of math instruction and alignment to the Common Core State Standards, common assessments for all subjects and grade levels and professional learning communities. From 2010–2013, Sarah served as a member and chair of the National Council of Teachers of Mathematics editorial panel for their Mathematics Teacher journal. Sarah earned a Masters of Science in Teaching Mathematics from Portland State University.

Michelle Terry has been a classroom teacher in the Estacada and Newberg School Districts in Oregon. Michelle has received training in Professional Learning Communities, Critical Friends, ELL Instructional Strategies, Proficiency-Based Grading and Lesson Design, Power Strategies for Effective Teaching, and Classroom Love and Logic. Michelle has an Interdisciplinary Masters from Western Oregon University. She currently teaches mathematics at Newberg High School.

Scott Valway has been a classroom teacher in the Tigard-Tualatin, Newberg and Parkrose School Districts in Oregon. Scott has been trained in Differentiated Instruction, Professional Learning Communities, Critical Friends, Discovering Algebra, Pre-Advanced Placement, Assessment Writing and Credit by Proficiency. Scott has a Masters of Science in Teaching from Oregon State University. He currently teaches math at Parkrose High School.

COMMON CORE STATE STANDARDS
Grade 7 Overview

The complete set of Common Core State Standards can be found at http://www.corestandards.org/. This book focuses on the highlighted Common Core State Standards shown below.

Ratios and Proportional Relationships

- Analyze proportional relationships and use them to solve real-world and mathematical problems.

The Number System

- Apply and extend previous understanding of operations with fractions to add, subtract, multiply and divide rational numbers.

Expressions and Equations

- Use properties of operations to generate equivalent expressions.

- Solve real-life and mathematical problems using numerical and algebraic expressions and equations.

Geometry

- ==Draw, construct and describe geometrical figures and describe the relationships between them.==

- ==Solve real-life and mathematical problems involving angle measure, area, surface area and volume.==

Statistics and Probability

- Use random sampling to draw inferences about a population.

- Draw informal comparative inferences about two populations.

- Investigate chance processes and develop, use and evaluate probability models.

Mathematical Practices

1. Make sense of problems and persevere in solving them.

2. Reason abstractly and quantitatively.

3. Construct viable arguments and critique the reasoning of others.

4. Model with mathematics.

5. Use appropriate tools strategically.

6. Attend to precision.

7. Look for and make use of structure.

8. Look for and express regularity in repeated reasoning.

CORE FOCUS ON SHAPES & ANGLES
CONTENTS IN BRIEF

How To Use Your Math Book	VIII
Block 1 Angle Relationships	1
Block 2 Two-Dimensional Geometry	35
Block 3 Surface Area and Volume	85
Acknowledgements	125
English/Spanish Glossary	127
Selected Answers	161
Index	164
Problem-Solving	166
Symbols	167

CORE FOCUS ON SHAPES & ANGLES

BLOCK 1 ~ ANGLE RELATIONSHIPS

Lesson 1.1	Measuring and Naming Angles	3
Lesson 1.2	Classifying Angles	8
	Explore! Classify an Angle	
Lesson 1.3	Complementary and Supplementary Angles	13
	Explore! Complementary vs. Supplementary	
Lesson 1.4	Vertical Angles and Adjacent Angles	19
	Explore! The Vertical Angle Relationship	
Lesson 1.5	Drawing Geometric Shapes	24
	Explore! Knowing Three Measures	
Review	Block 1 ~ Angle Relationships	30

BLOCK 2 ~ TWO-DIMENSIONAL GEOMETRY

Lesson 2.1	Areas of Triangles and Parallelograms	37
Lesson 2.2	Area of a Trapezoid	42
	Explore! A Formula for Trapezoid Area	
Lesson 2.3	Parts of a Circle	47
Lesson 2.4	Circumference and Pi	52
	Explore! A Special Ratio	
Lesson 2.5	Area of a Circle	56
	Explore! Circle Areas	
Lesson 2.6	More Pi	61
	Explore! Which Pi?	
Lesson 2.7	Composite Figures	65
Lesson 2.8	Circle Similarity	69
	Explore! Stepping Stones	
Lesson 2.9	Area of Sectors	74
Review	Block 2 ~ Two-Dimensional Geometry	78

BLOCK 3 ~ SURFACE AREA AND VOLUME

Lesson 3.1	Three-Dimensional Figures	87
Lesson 3.2	Drawing Solids	92
	Explore! Netting A Solid	
Lesson 3.3	Slicing Solids	96
	Explore! Cutting Clay	
Lesson 3.4	Surface Area of Prisms	101
	Explore! Take Your Pick	
Lesson 3.5	Volume of Prisms	106
	Explore! Cutting Corners	
Lesson 3.6	Surface Area of Regular Pyramids	111
	Explore! Tent Making	
Lesson 3.7	Volume of Pyramids	115
	Explore! Pyramid vs. Prism	
Review	Block 3 ~ Surface Area and Volume	119

HOW TO USE YOUR MATH BOOK

Your math book has features that will help you be successful in this course. Use this guide to help you understand how to use this book.

Lesson Target

 Look in this box at the beginning of every lesson to know what you will be learning about in each lesson.

Vocabulary

Each new vocabulary word is printed in **red**. The definition can be found with the word. You can also find the definition of the word in the glossary which is in the back of this book.

Explore!

Some lessons have **EXPLORE!** activities which allow you to discover mathematical concepts. Look for these activities in the Table of Contents and in lessons next to the purple line.

Examples

Examples are useful because they remind you how to work through different types of problems. Look for the word **EXAMPLE** and the green line.

Helpful Hints

Helpful hints and important things to remember can be found in green callout boxes.

Blue Boxes

A blue box holds important information or a process that will be used in that lesson. Not every lesson has a blue box.

 This calculator icon will appear in Lessons and Exercises where a calculator is needed. Your teacher may want you to use your calculator at other times, too. If you are unsure, make sure to ask if it is the right time to use it.

EXERCISES

The **EXERCISES** are a place for you to find practice problems to determine if you understand the lesson's target. You can find selected answers in the back of this book so you can check your progress.

REVIEW

The **REVIEW** provides a set of problems for you to practice concepts you have already learned in this book. The **REVIEW** follows the **EXERCISES** in each lesson. There is also a **REVIEW** section at the end of each Block.

TIC-TAC-TOE ACTIVITIES

Each Block has a Tic-Tac-Toe board at the beginning with activities that extend beyond the Common Core State Standards. The Tic-Tac-Toe activities described on the board can be found throughout each Block in yellow boxes.

CAREER FOCUS

At the end of each Block, you will find an autobiography of an individual. Each one describes what they like about their job and how math is used in their career.

CORE FOCUS ON MATH
STAGE 2

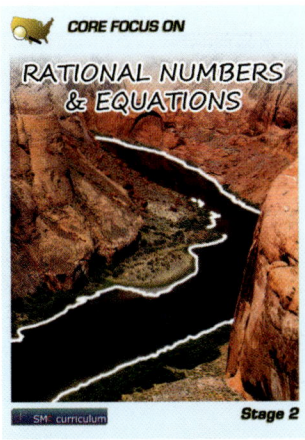 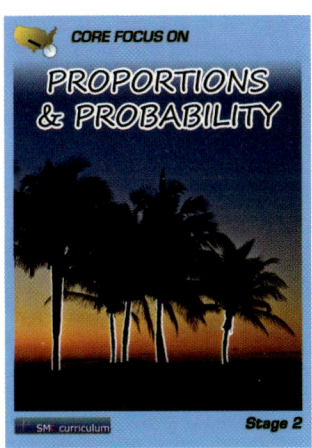 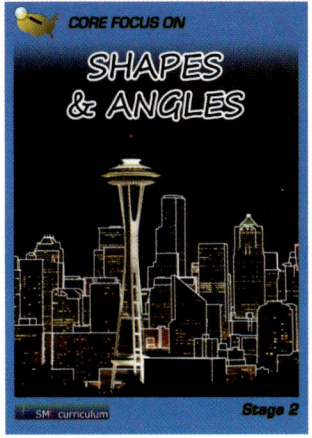

CORE FOCUS ON SHAPES & ANGLES
BLOCK 1 ~ ANGLE RELATIONSHIPS

Lesson 1.1	Measuring and Naming Angles	3
Lesson 1.2	Classifying Angles	8
	Explore! Classify an Angle	
Lesson 1.3	Complementary and Supplementary Angles	13
	Explore! Complementary vs. Supplementary	
Lesson 1.4	Vertical Angles and Adjacent Angles	19
	Explore! The Vertical Angle Relationship	
Lesson 1.5	Drawing Geometric Shapes	24
	Explore! Knowing Three Measures	
Review	Block 1 ~ Angle Relationships	30

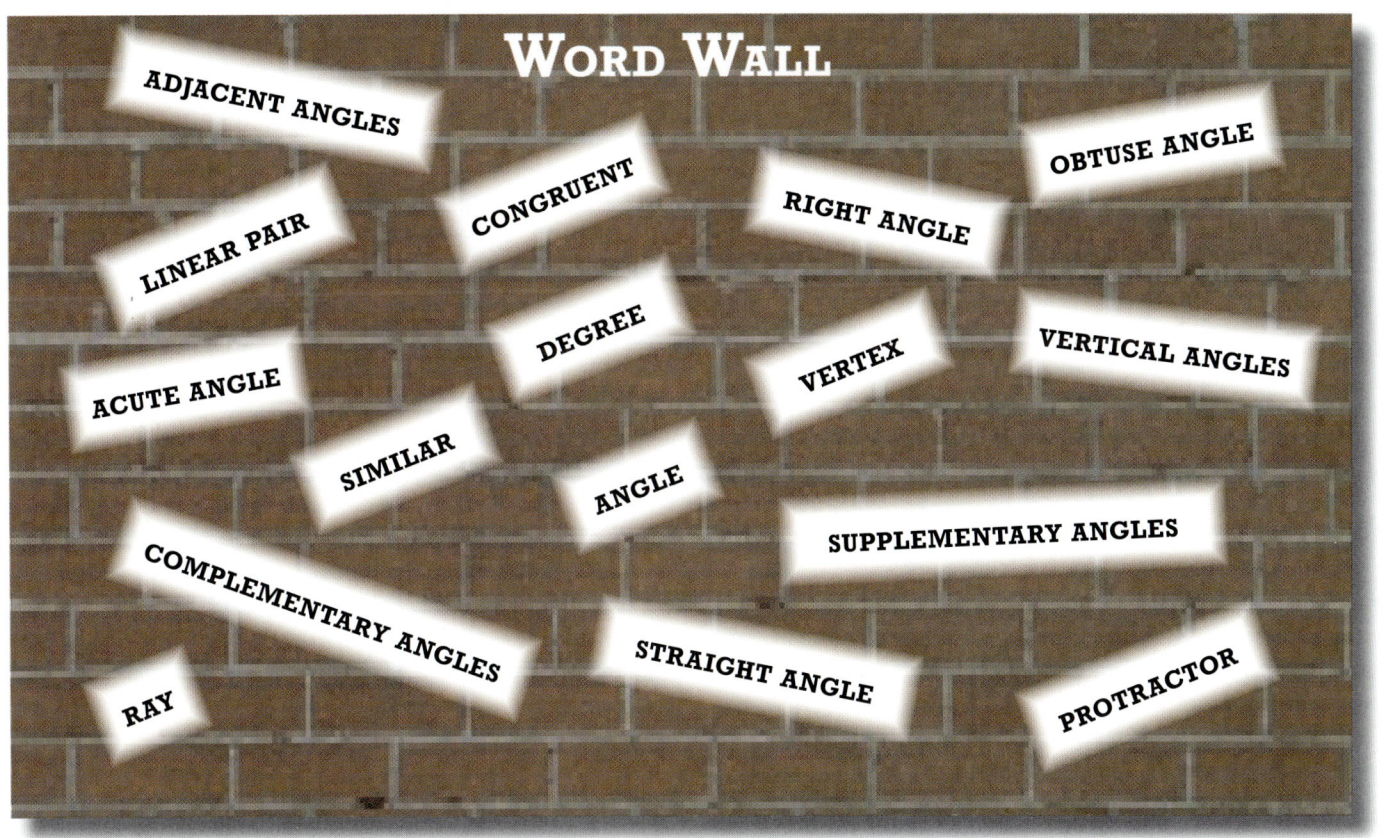

BLOCK 1 ~ ANGLE RELATIONSHIPS TIC-TAC-TOE

PUZZLING ANGLES Find angle measures in a complex diagram.  *See page 29 for details.*	**BISECTING ANGLES** Use two types of constructions to bisect angles. 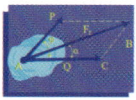 *See page 18 for details.*	**A TRANSVERSAL OF ANGLES** Create a flip book which describes special angle pairs.  *See page 23 for details.*
CROSSWORD Make a crossword using vocabulary from this block.  *See page 29 for details.*	**PROTRACTOR GUIDE** Write a guide for using a protractor. 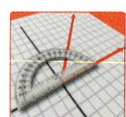 *See page 12 for details.*	**TRIANGLE COLLAGE** Find or take pictures of triangles and display them.  *See page 33 for details.*
ANGLE ART Create an original piece of artwork with lines and angles.  *See page 17 for details.*	**HIDDEN TRIANGLES** Count triangles and create a triangle design.  *See page 33 for details.*	**DUPLICATING ANGLES** Use a compass and straightedge to duplicate angles.  *See page 7 for details.*

MEASURING AND NAMING ANGLES

LESSON 1.1

 Measure, name and draw angles.

Angles are used in construction, architecture, graphic design, aerospace, art, machining and manufacturing, as well as many other fields. An **angle** is formed by two rays with a common endpoint. A **ray** has one endpoint and extends forever in one direction.

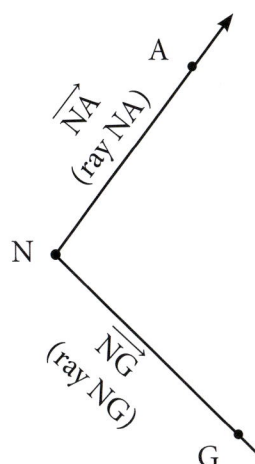

Ray NA is written $\overrightarrow{NA}$. Ray NG is written $\overrightarrow{NG}$. The first point in the name of a ray is the endpoint of the ray.

The **vertex** of an angle is the common point of both rays. N is the vertex of this angle.

When three points are used to name an angle, the vertex is written in the middle of the name. The vertex can be written as the name of an angle when it is the vertex for only one angle.

Three ways to name the angle formed by $\overrightarrow{NA}$ and $\overrightarrow{NG}$ are ∠ANG, ∠GNA and ∠N.

EXAMPLE 1 Give 4 different names for the given angle.

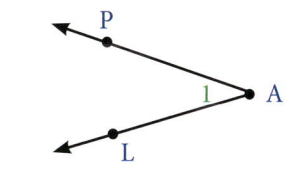

An angle can also be named using a number.

SOLUTIONS 1. ∠PAL 2. ∠LAP 3. ∠A 4. ∠1

EXAMPLE 2 Is ∠W another name for ∠NWE? Explain your reasoning.

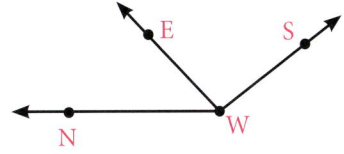

SOLUTION No, it is not clear whether ∠W refers to ∠NWE, ∠SWE or ∠NWS.

Adjacent angles are two angles that share a ray. In **Example 2**, ∠NWE and ∠SWE share $\overrightarrow{WE}$. This means ∠NWE and ∠SWE are adjacent angles.

EXAMPLE 3 Identify at least one additional name for each angle. Write using proper angle notation.

a. ∠1

b. ∠TGI

c. ∠RGE

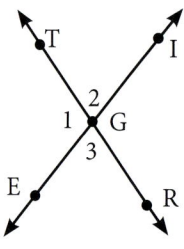

SOLUTIONS

a. ∠TGE or ∠EGT

b. ∠2 or ∠IGT

c. ∠EGR or ∠3

A **protractor** is a tool used to measure angles. Angles are measured in units called **degrees**.

USING A PROTRACTOR TO MEASURE ANGLES

1. Place the bottom edge of the protractor on one ray of the angle.
2. Slide the protractor so the center of the bottom edge is on the vertex of the angle.
3. Begin at 0° and use your finger to curve around the protractor until your finger is lined up with the other ray of the angle. The number that the ray lines up with is the measure of the angle.

The "*m*" in front of an angle measure is notation for the word "measure".
The statement in **Figure 1** below reads, "The measure of ∠ABC is equal to sixty degrees."

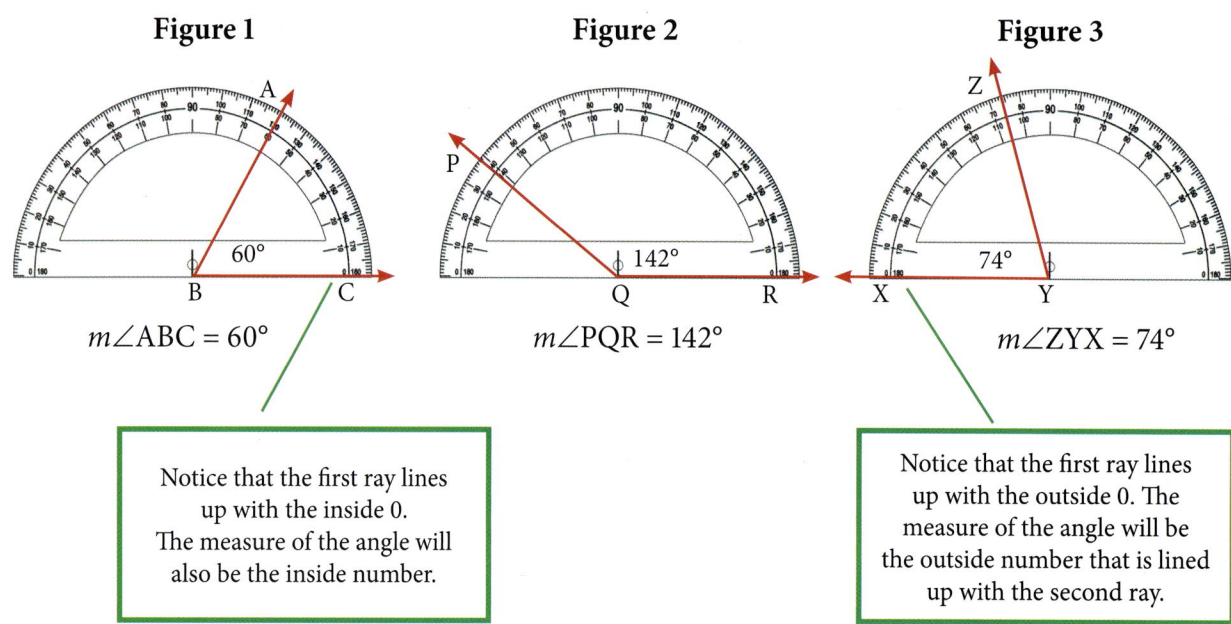

Notice that the first ray lines up with the inside 0. The measure of the angle will also be the inside number.

Notice that the first ray lines up with the outside 0. The measure of the angle will be the outside number that is lined up with the second ray.

Lesson 1.1 ~ Measuring and Naming Angles

EXAMPLE 4 Use a protractor to measure each angle.

a. 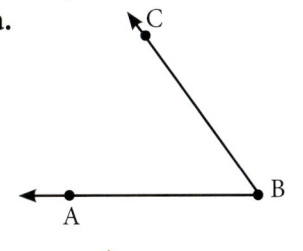   b.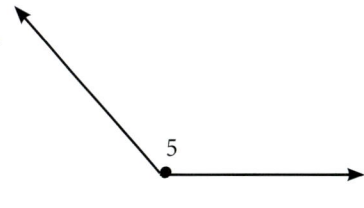

SOLUTIONS a.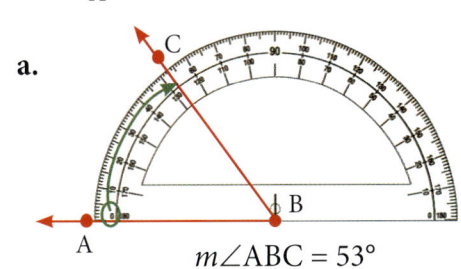

$m\angle ABC = 53°$

b.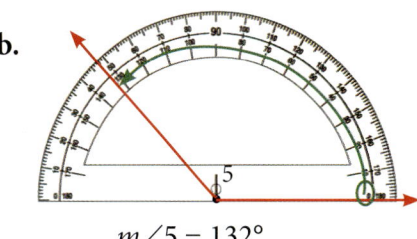

$m\angle 5 = 132°$

EXERCISES

Give two different names for each angle.

1. 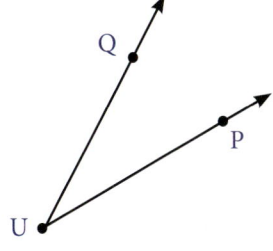   2. 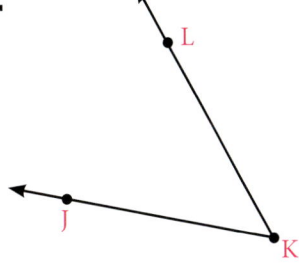   3.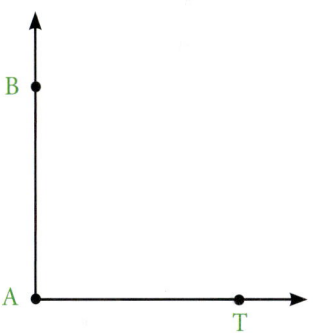

Sketch a diagram to represent each angle.

4. ∠DOG

5. ∠CUB also called ∠4

6. ∠PAL also called ∠2

7. ∠1 and ∠2 are adjacent angles

8. ∠XYZ and ∠XYU are adjacent angles

9. ∠HOT is approximately 90°

Use each protractor to determine the measure of the angle.

10. 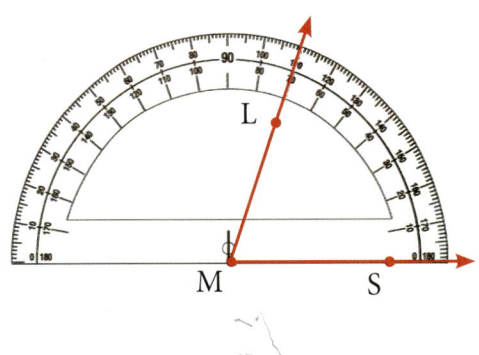   11.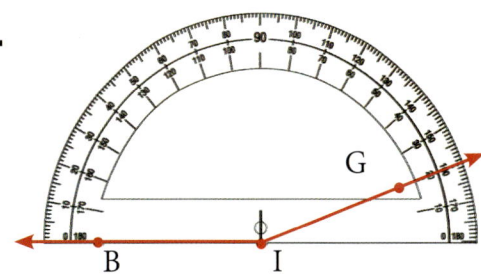

Lesson 1.1 ~ Measuring and Naming Angles 5

Use a protractor to measure each angle to the nearest degree.

12.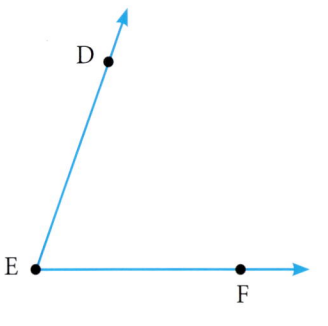

13.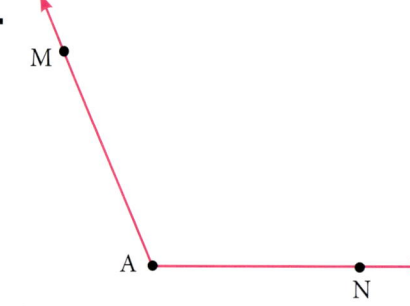

14.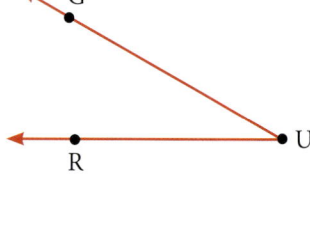

15.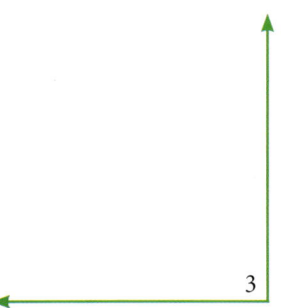

16.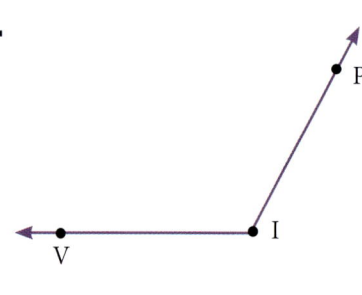

17.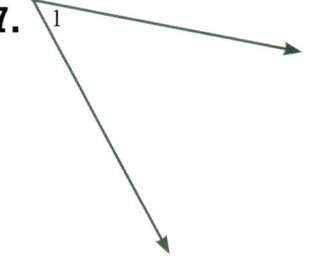

Use a protractor to draw each angle. Label the angle(s).

18. $m\angle SAM = 34°$

19. $m\angle YAK = 115°$

20. two 35° angles with the same vertex

21. two adjacent angles that are 50° and 100°

Use the diagram below to name an angle with the specified measure.

22. 90°

23. 50°

24. 17°

25. 145°

26. 163°

27. 130°

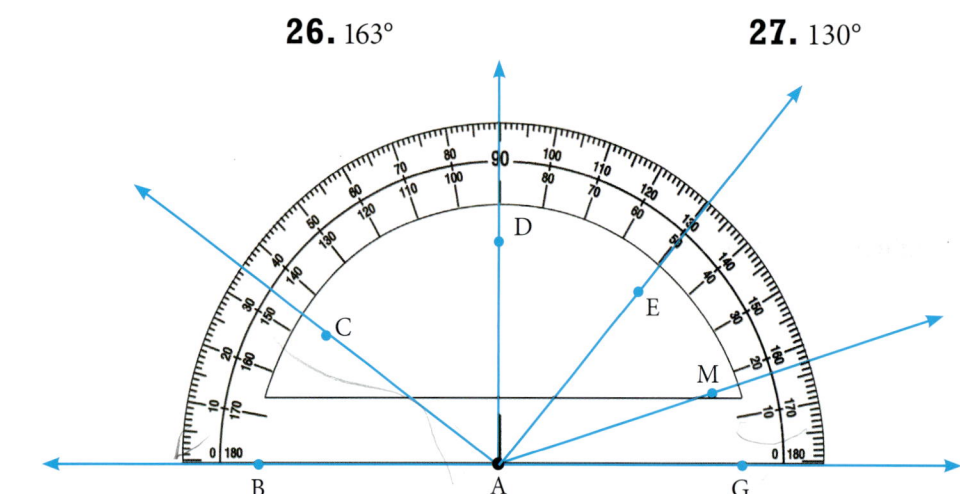

28. Marcus uses the protractor diagram above and says $m\angle EAD = 40°$. Is he correct? Show all work necessary to justify your answer.

29. $\angle PRM$ and $\angle MRT$ are adjacent angles. Will $m\angle PRT$ *always*, *sometimes* or *never* be greater than $m\angle PRM$? Explain how you know your answer is correct.

6   Lesson 1.1 ~ Measuring and Naming Angles

Tic-Tac-Toe ~ Duplicating Angles

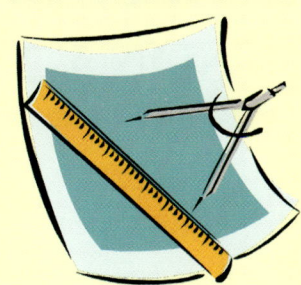

Constructions are part of geometry. A geometric construction is made by using a compass and straightedge. Follow the steps to duplicate ∠ABC on a piece of notebook paper.

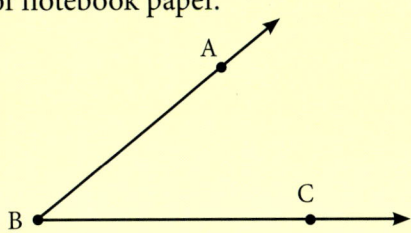

Step 1: Trace ∠ABC on your paper.

Step 2: Use a straightedge to draw a ray. This will be one side of the duplicate angle.

Step 3: Place the stylus or sharp point of a compass on the vertex of the traced angle. Draw an arc on the angle.

Step 4: Without changing the setting on the compass, place the stylus on the endpoint of the duplicate ray and draw an arc.

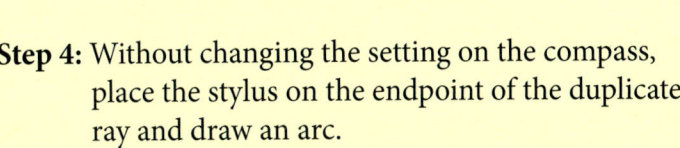

Step 5: Use the compass to measure the width of the arc drawn on the original angle. Place the stylus on the intersection of a side and the arc. Adjust the compass so the pencil is touching the other intersection point.

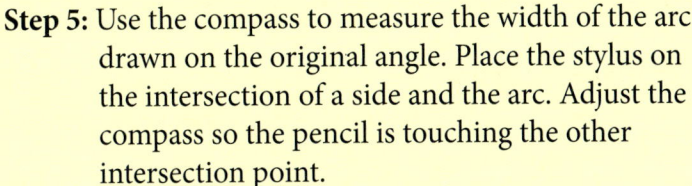

Step 6: Without changing the setting on the compass, place the stylus on the intersection of the arc and duplicate ray. Make a small arc intersecting the larger arc.

Step 7: Use a straightedge to connect the endpoint of the duplicate ray. This is the second ray needed to complete the duplication of ∠ABC.

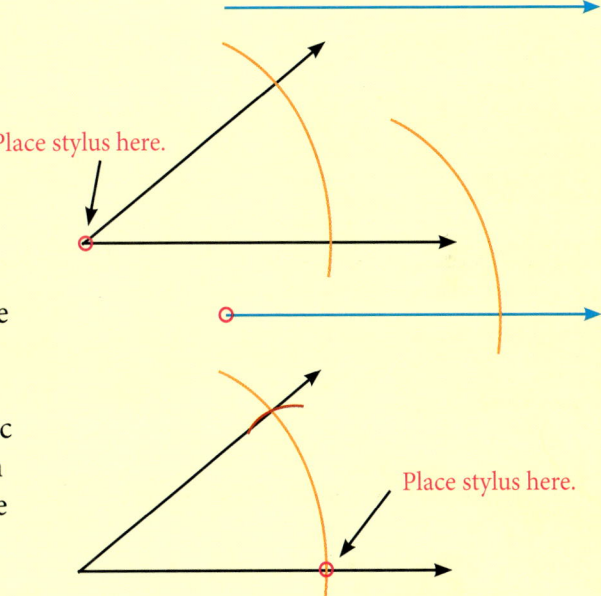

1. Use a protractor to draw a 60° angle.

2. Using only a compass and straightedge, duplicate the 60° angle.

3. Measure the duplicated angle with a protractor to check accuracy.

4. Repeat these steps on a 25°, 128° and 160° angle.

5. Which step is the most difficult for you in this process? Why?

Lesson 1.1 ~ Measuring and Naming Angles

CLASSIFYING ANGLES

LESSON 1.2

 Classify angles as acute, right, obtuse or straight.

In **Lesson 1.1** you named and measured angles. Angles can be classified into groups by their degree measure.

EXPLORE! **CLASSIFY AN ANGLE**

Step 1: Use a protractor to measure the angles in each group. Record each measurement.

GROUP A
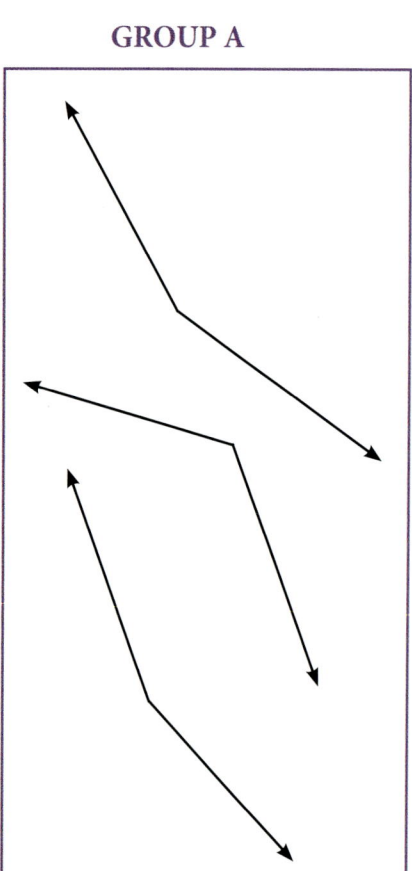

GROUP B
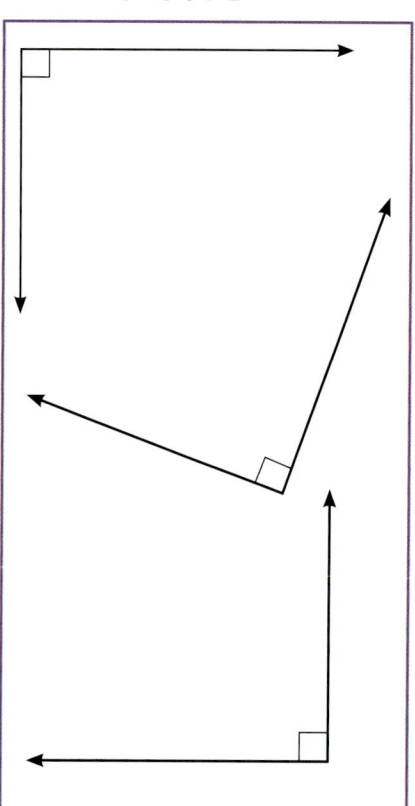

GROUP C
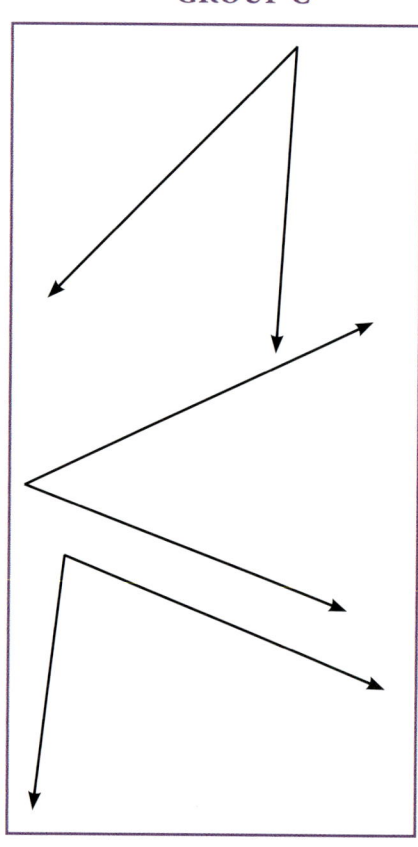

Step 2: Use the measures recorded in **Step 1** to answer each question for each group.
   a. How do the angles in Group A compare to Group B?
   b. How do the angles in Group C compare to Group B?

Step 3: Write at least two sentences describing each group of angles. Include information about their degree measures.

Step 4: Compare what you wrote in **Step 3** to the definitions given below. Identify which definition is appropriate for each group.
   Acute angle – an angle that measures more than 0° and less than 90°
   Right angle – an angle that measures exactly 90°
   Obtuse angle – an angle that measures more than 90° but less than 180°

There are four classifications of angles based on their degree measure. In the **Explore!**, you learned about acute, obtuse and right angles. An angle that has a measure of 180° is called a **straight angle**.

Right angles are often identified by drawing a small square in the vertex of the angle. If a square is present in the vertex of an angle, the angle measures 90°.

Indicates right angle (90°)

EXAMPLE 1 Classify each angle by measuring it with a protractor.

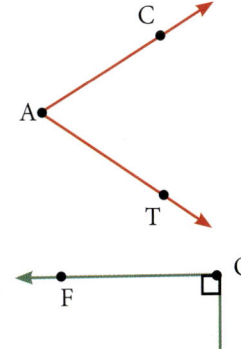
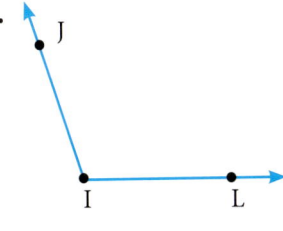

SOLUTIONS
a. Acute, $m\angle CAT = 64°$ which is less than 90°.
b. Obtuse, $m\angle JIL = 110°$ which is more than 90°.
c. Right, $m\angle FOX = 90°$ which is shown by placing a square in the vertex.
d. Straight, $m\angle DOG = 180°$.

Angles with equal measures are **congruent**. The symbol for congruent is ≅. Congruent angles are identified in diagrams with congruence marks. The two marks on the arc inside each angle below are congruence marks. They show ∠RED ≅ ∠BLU. This is read, "Angle RED is congruent to angle BLU."

∠X is congruent to ∠Y because each measures 90°. Each angle is a right angle as indicated by the square in each vertex.

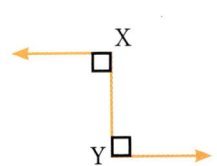
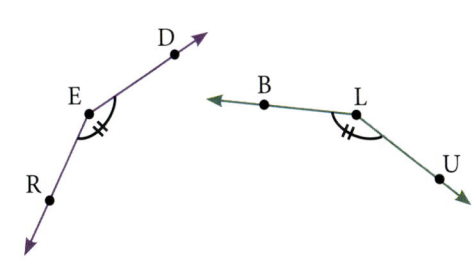

Lesson 1.2 ~ Classifying Angles **9**

EXAMPLE 2 Sketch a diagram of two angles that are congruent and adjacent.

SOLUTION The angles share $\overrightarrow{OC}$ which makes them adjacent. The congruence marks indicate that ∠DOC is congruent to ∠BOC.

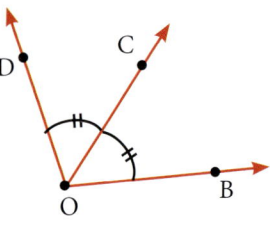

EXAMPLE 3 Use the information in the diagram to write an equation. Solve for x.

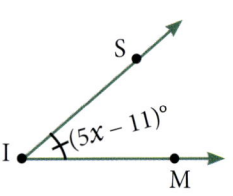
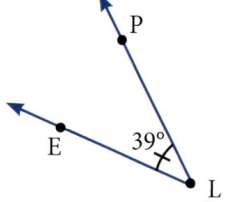

SOLUTION Congruence marks show the angles are congruent. ∠SIM ≅ ∠PLE
Write the equation showing degree measures are equal. $5x - 11 = 39$
Add 11 to both sides of the equation. $+11 \quad +11$
Divide both sides of the equation by 5. $\dfrac{5x}{5} = \dfrac{50}{5}$

$x = 10$

☑ Check the solution by substituting 10 for x.

$$5x - 11 = 39$$
$$5(10) - 11 \stackrel{?}{=} 39$$
$$50 - 11 \stackrel{?}{=} 39$$
$$39 = 39$$

EXAMPLE 4 ∠JAK is congruent to ∠HIL. The measure of ∠JAK = $(12 - 3x)°$ and the measure of ∠HIL = $(44 - x)°$. Solve for x. Then find the measure of each angle.

SOLUTION Write an equation showing the angles ∠JAK ≅ ∠HIL
have equal measures. $12 - 3x = 44 - x$
Add x to each side of the equation. $+x \quad\quad +x$
  $12 - 2x = 44$
Subtract 12 from each side of the equation. $-12 \quad\quad -12$
Divide each side of the equation by −2. $\dfrac{-2x}{-2} = \dfrac{32}{-2}$

$x = -16$

Write the given expression
for each angle. $m∠JAK = 12 - 3x$ $\quad$ $m∠HIL = 44 - x$
Substitute −16 for x. $= 12 - 3(-16)$ $\quad$ $= 44 - (-16)$
Simplify. $= 12 + 48$ $\quad\quad\quad$ $= 44 + 16$
Add. $= 60$ $\quad\quad\quad\quad\quad$ $= 60$

$x = -16$ which means $m∠JAK = 60°$ and $m∠HIL = 60°$. Both angles are equal which verifies that they are congruent.

10 Lesson 1.2 ~ Classifying Angles

EXERCISES

Estimate the degree measure for each angle. Classify each angle as acute, obtuse, right or straight.

1.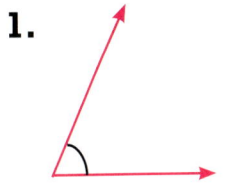

2.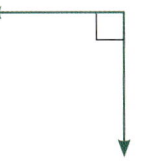

3.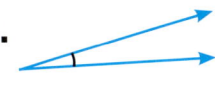

4.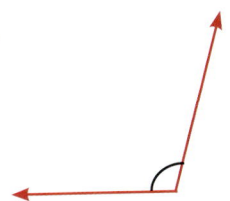

5.

6.

Sketch a diagram for each description. Label each angle.

7. two congruent angles

8. ∠GUY is obtuse

9. ∠GAL is acute

10. ∠JKL ≅ ∠POM

11. a right angle that can be identified using 3 names

12. two adjacent, right angles

13. two congruent angles that share a vertex

14. ∠PQR is straight

Use the given information to solve for *x*.

15.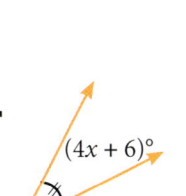

16. $m\angle PQR = 124°$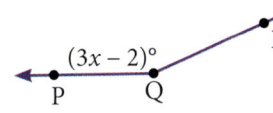

17.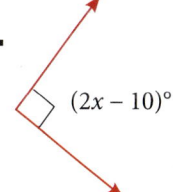

18.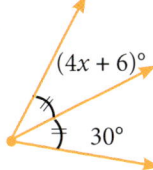

19. ∠ROY ≅ ∠MAN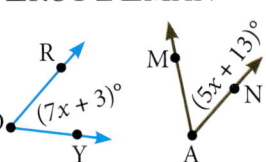

20. $m\angle XYZ = m\angle ABC$
$m\angle XYZ = (5 + 2x)°$
$m\angle ABC = (3x + 1)°$

21. ∠JAM and ∠GEM are congruent angles. The measure of ∠JAM is $(8x + 5)°$ and $m\angle GEM = (x + 75)°$. Find the measure of each angle. Show all work necessary to justify your answer.

22. ∠SML is an acute angle. The measure of ∠SML = $(x - 7)°$. Which values must *x* be between? Use words and/or numbers to show how you determined your answer.

Lesson 1.2 ~ Classifying Angles **11**

23. $m\angle LRG = (4x + 22)°$

   a. If $\angle LRG$ is a right angle, what must x equal?

   b. In order for $\angle LRG$ to be an acute angle, Latoya says the value of x must be between −5.5 and 17. Javier says the value of x must be between 0 and 17. Which person is correct? Explain your reasoning.

   c. If $\angle LRG$ is an obtuse angle, what inequality represents the possible values for x?

REVIEW

Sketch a diagram to represent each description.

24. $\angle RMP$

25. $\angle PIE$ also called $\angle 3$

26. $\angle 5$ and $\angle 6$ are adjacent

Use a protractor to measure each angle.

27.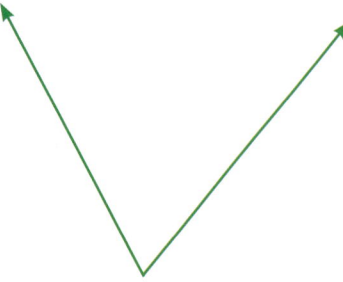

28.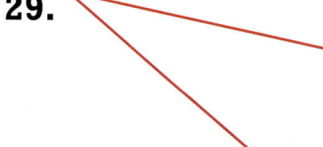

29.

Use a protractor to draw an angle with the given measure.

30. 67°

31. 135°

32. 180°

33. 11°

Tic-Tac-Toe ~ Protractor Guide

Step 1: Create a user's guide describing how to measure and draw angles with a protractor. Your teacher may use the guide to refresh a substitute teacher or for students who are absent the day of the lesson.

Step 2: Create a worksheet that can be completed using the guide.

Step 3: Make an answer key for the worksheet.

COMPLEMENTARY AND SUPPLEMENTARY ANGLES

LESSON 1.3

Use facts about complementary and supplementary angles to solve problems.

Individual angles are classified as acute, right, obtuse or straight. Special pairs of angles can also be classified. In this lesson two special pairs of angles and their relationships will be examined.

EXPLORE!   COMPLEMENTARY VS SUPPLEMENTARY

Step 1: Look at the angles in the chart.
  a. What similarities do you notice about the pairs of angles called supplementary angles?
  b. What similarities do you notice about the pairs of angles called complementary angles?

SUPPLEMENTARY ANGLES		COMPLEMENTARY ANGLES	
(figure: angles 1 and 2 on a line)	∠1 and ∠2 are supplementary.	*(figure: right angle ROE with ray OM)*	∠ROM and ∠MOE are complementary.
(figure: ∠CAT = 40° and ∠DOG = 140°)	∠CAT and ∠DOG are supplementary.	*(figure: ∠CAR = 70° and ∠WLK = 20°)*	∠CAR and ∠WLK are complementary.
(figure: right angles LEF and RGT)	∠LEF and ∠RGT are supplementary.	*(figure: angles 1 and 2 forming a right angle)*	∠1 and ∠2 are complementary.
m∠6 = 80° and m∠7 = 100°	∠6 and ∠7 are supplementary.	m∠8 = 45° and m∠9 = 45°	∠8 and ∠9 are complementary.

Step 2: Write a definition for supplementary angles.

Step 3: Write a definition for complementary angles.

Step 4: Give at least two examples of angle measures for each type of angle pair listed below.
  a. Supplementary angles
  b. Complementary angles

Lesson 1.3 ~ Complementary and Supplementary Angles

Complementary and supplementary angles are special pairs of angles. **Complementary angles** are two angles with a sum of 90°. Two angles with a sum of 180° are called **supplementary angles**. These special pairs of angles may or may not be adjacent (share a side).

EXAMPLE 1 Use the diagram to find $m\angle PAR$.

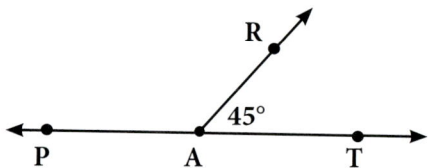

SOLUTION

$\angle PAR$ and $\angle TAR$ together form a straight angle, $\angle PAT$. They are supplementary angles.

Supplementary angles have a sum of 180°.
Subtract 45 from both sides of the equation.

$m\angle PAR + 45° = 180°$
$ -45° \quad -45°$
$m\angle PAR = 135°$

✓ Check the solution.

$135° + 45° \stackrel{?}{=} 180°$
$180° = 180°$

$m\angle PAR = 135°$

EXAMPLE 2 $\angle GRA$ and $\angle INS$ are supplementary.
a. Write an equation and solve for x.
b. Determine the measure of each angle.

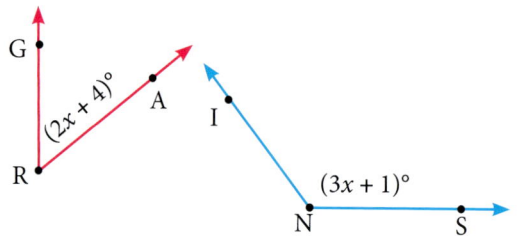

SOLUTIONS

a. Supplementary angles have a sum of 180°.
Write an equation.

$m\angle GRA + m\angle INS = 180°$
$(2x + 4) + (3x + 1) = 180$

Combine like terms.
Subtract 5 from each side.
Divide each side by 5.

$5x + 5 = 180$
$ -5 \quad -5$
$\dfrac{5x}{5} = \dfrac{175}{5}$
$x = 35$

b. Write the given expression for each angle.
Substitute 35 for x.
Multiply.
Add.

$m\angle GRA = (2x + 4)°$
$= (2(35) + 4)°$
$= (70 + 4)°$
$= 74°$

$m\angle INS = (3x + 1)°$
$= (3(35) + 1)°$
$= (105 + 1)°$
$= 106°$

✓ $m\angle GRA + m\angle INS = 180°$
$74° + 106° \stackrel{?}{=} 180°$
$180° = 180°$

The measure of $\angle GRA$ is 74°.
The measure of $\angle INS$ is 106°.

Lesson 1.3 ~ Complementary and Supplementary Angles

EXAMPLE 3

Use the diagram to write an equation. Solve for x.

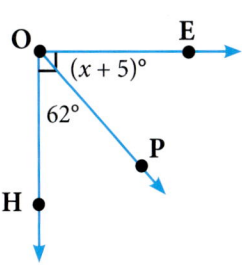

SOLUTION

Complementary angles have a sum of 90°. $m\angle HOP + m\angle EOP = 90°$
Substitute the degree measures. $62 + (x + 5) = 90$
Combine the like terms. $x + 67 = 90$
Subtract 67 from each side of the equation. $\underline{-67 -67}$
$x = 23°$

☑ Check the solution. $62 + (23 + 5) \stackrel{?}{=} 90$
$62 + 28 \stackrel{?}{=} 90$
$90 = 90$

The value of x is 23.

EXAMPLE 4

$\angle 1$ and $\angle 2$ are complementary angles. The measure of $\angle 1 = (3x + 4)°$ and $m\angle 2 = (x + 6)°$.
a. Draw a diagram.
b. Write an equation and solve for x.
c. Find $m\angle 1$ and $m\angle 2$.

SOLUTIONS

a. 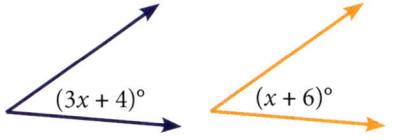 or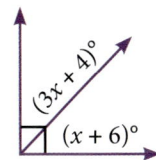

b. Complementary angles have a sum of 90°. $m\angle 1 + m\angle 2 = 90°$
Substitute the degree measures. $(3x + 4) + (x + 6) = 90$
Combine like terms. $4x + 10 = 90$
Subtract 10 from each side of the equation. $\underline{-10 -10}$
Divide by 4 on each side of the equation. $\dfrac{4x}{4} = \dfrac{80}{4}$
$x = 20$

c. Write the given expression
for each angle. $m\angle 1 = (3x + 4)°$ $m\angle 2 = (x + 6)°$
Substitute 20 for x. $= (3(20) + 4)$ $= (20 + 6)$
Multiply. $= (60 + 4)$
Add. $= 64°$ $= 26°$

☑ Check the solution. $64° + 26° \stackrel{?}{=} 90°$
$90 = 90$

$m\angle 1 = 64°$ and $m\angle 2 = 26°$.

EXERCISES

Identify each pair of angles as complementary, supplementary or neither.

1.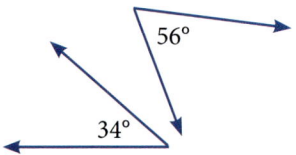

2.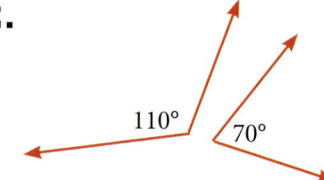

3.

4.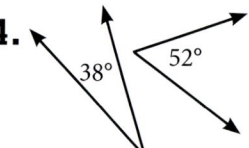

5.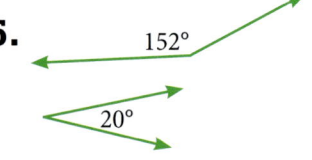

6.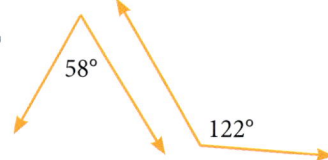

7. $m\angle 1$ and $m\angle 2$ sum to $181°$.

8. $\angle A$ and $\angle M$ have a sum of $90°$.

Write an equation for each description. Solve for *x*. Check your solution.

9. $\angle A$ and $\angle B$ are complementary

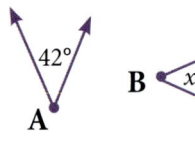

10.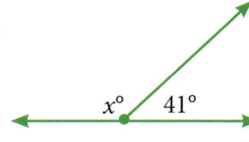

11.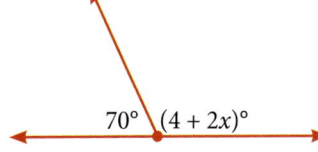

12.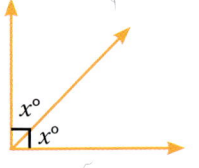

13.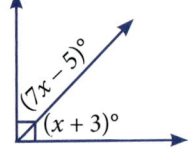

14.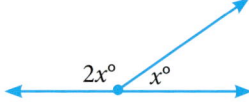

Show all work necessary to justify each answer.

15. $\angle MAN$ and $\angle MAP$ are supplementary. The measure of $\angle MAN$ is $57°$. What is the measure of $\angle MAP$?

16. $\angle 5$ and $\angle 7$ are complementary angles. Find the measure of $\angle 7$ if $m\angle 5 = 47°$.

17. The complement of ∠Q is 31°. Find *m*∠Q.

18. The supplement of ∠U is 62°. What is *m*∠U?

Find the measure of each angle in Exercises 19–23. Use mathematics to justify each answer.

19. ∠1 and ∠2 are supplementary; *m*∠1 = 3x° and *m*∠2 = 3x°.

20. ∠W and ∠C are complementary; *m*∠W = (47 + 3x)° and *m*∠C = (10 + 8x)°.

21. ∠G and ∠H are supplementary; *m*∠G = (x + 4)° and *m*∠H = (4x + 11)°.

22. ∠V and ∠W are supplementary; *m*∠V = (12 + 3x)° and *m*∠W = (33 + 2x)°.

23. ∠1 and ∠2 are complementary; *m*∠1 = 4x° and *m*∠2 = (x + 8)°.

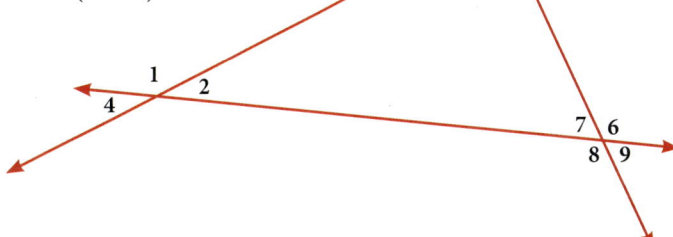

24. Use the figure on the right.
   a. Name an angle that appears to be obtuse.
   b. Name three pairs of supplementary angles.
   c. Name a pair of complementary angles.
   d. Give possible measures for ∠4 and ∠1.
   e. Determine *m*∠11.

25. Are two right angles *always*, *sometimes* or *never* supplementary? Explain your reasoning.

26. Are complementary angles *always*, *sometimes* or *never* adjacent? Explain your reasoning.

REVIEW

Draw and label a diagram to represent each statement.

27. ∠PAM is acute

28. ∠BIG is obtuse

29. ∠RHT is right

30. ∠ABC is an acute angle. The measure of ∠ABC = (x + 15)°. Write an inequality that represents the possible values for x.

31. ∠JKL and ∠XYZ are congruent angles. *m*∠JKL = (4x − 10)° and *m*∠XYZ = (3x + 15)°. Find the measure of each angle.

TIC-TAC-TOE ~ ANGLE ART

Step 1: Research how angles and lines are used in art. Write a summary of your findings that is at least one page in length.

Step 2: Create a work of art using the different types of angles in **Block 1**. Use an 11 inch by 18 inch piece of paper for your work of art. Give your creation a title and sign it.

Tic-Tac-Toe ~ Bisecting Angles

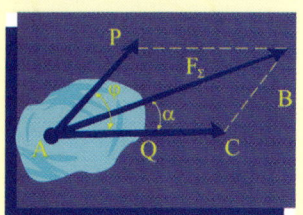

To bisect an angle means to cut it in half. You can bisect an angle using two different methods. One construction method to bisect an angle uses patty paper. Another method uses a compass and straightedge.

Bisecting an Angle Using Patty Paper or Tracing Paper

Step 1: Draw or trace an angle onto a piece of patty paper or tracing paper. Record the measure of the angle.

Step 2: Fold one ray of the angle onto the other ray of the angle.

Step 3: Trace the crease and measure each angle.

Bisecting an Angle Using A Compass and Straightedge

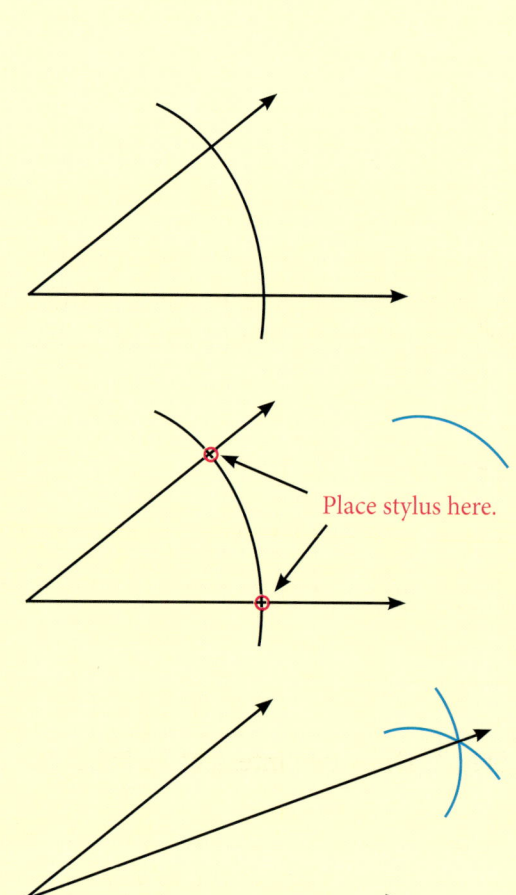

Step 1: Draw or trace an angle. Record the measure of the angle.

Step 2: Place the stylus or sharp point of a compass on the vertex. Use the compass to draw an arc through the angle.

Step 3: Place the stylus on one of the points of intersection between a ray of the angle and the arc from **Step 2.** Draw an arc as shown.

Place stylus here.

Step 4: Without changing the setting on the compass, repeat **Step 3** at the other point of intersection.

Step 5: Use a straightedge to draw a ray from the vertex to the intersection of the two arcs drawn in **Steps 3 and 4**.

Step 6: Measure each angle.

1. Use a protractor and patty/tracing paper to draw 90°, 24°, 115° and 160° angles. Each angle should be on a separate sheet of patty/tracing paper.
2. Bisect each angle by folding.
3. Use a protractor to draw another set of angles with the measures listed in **# 1** on regular paper.
4. Bisect each angle using a compass and straightedge.
5. Lay your matching patty/tracing paper constructions on top of each compass construction.
6. Summarize each method of construction. Include which method you prefer and explain why. Discuss the pros and cons of each method.

VERTICAL ANGLES AND ADJACENT ANGLES

LESSON 1.4

 Use facts about vertical and adjacent angles to solve problems.

Complementary and supplementary angles are types of special angles. Another pair of special angles are **vertical angles**. Vertical angles are formed by two intersecting lines. They have a common vertex but are not adjacent.

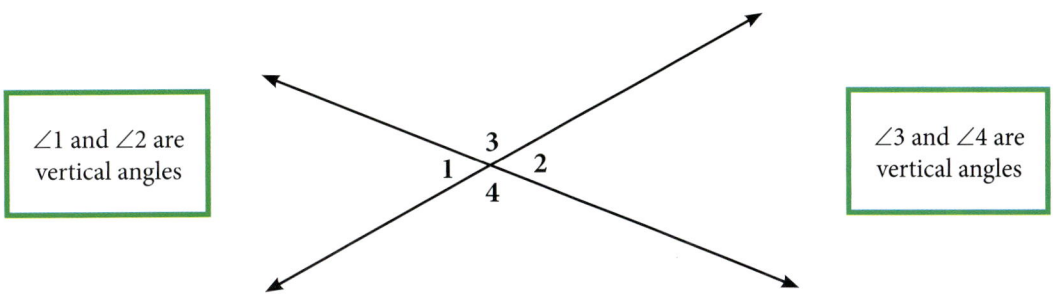

∠1 and ∠2 are vertical angles

∠3 and ∠4 are vertical angles

Two adjacent angles whose non-common sides are opposite rays are a **linear pair**. In the diagram above, there are four sets of linear pairs: ∠1 and ∠3, ∠1 and ∠4, ∠4 and ∠2 and ∠3 and ∠2. Notice each pair of these angles together form a straight angle. Each linear pair is supplementary.

EXPLORE!

Step 1: Trace ∠3 above onto a sheet of paper.

Step 2: Place the traced ∠3 on top of ∠4. What do you notice?

Step 3: Repeat **Steps 1 and 2** with ∠1 and ∠2. What do you notice?

Step 4: Draw two intersecting lines on a piece of paper.

Step 5: Label the angles that are formed using the numbers 5, 6, 7 and 8. Identify the vertical angles in your drawing.

Step 6: Measure each angle in your drawing with a protractor.

Step 7: What can you conclude about the measure of any pair of vertical angles?

THE VERTICAL ANGLE RELATIONSHIP

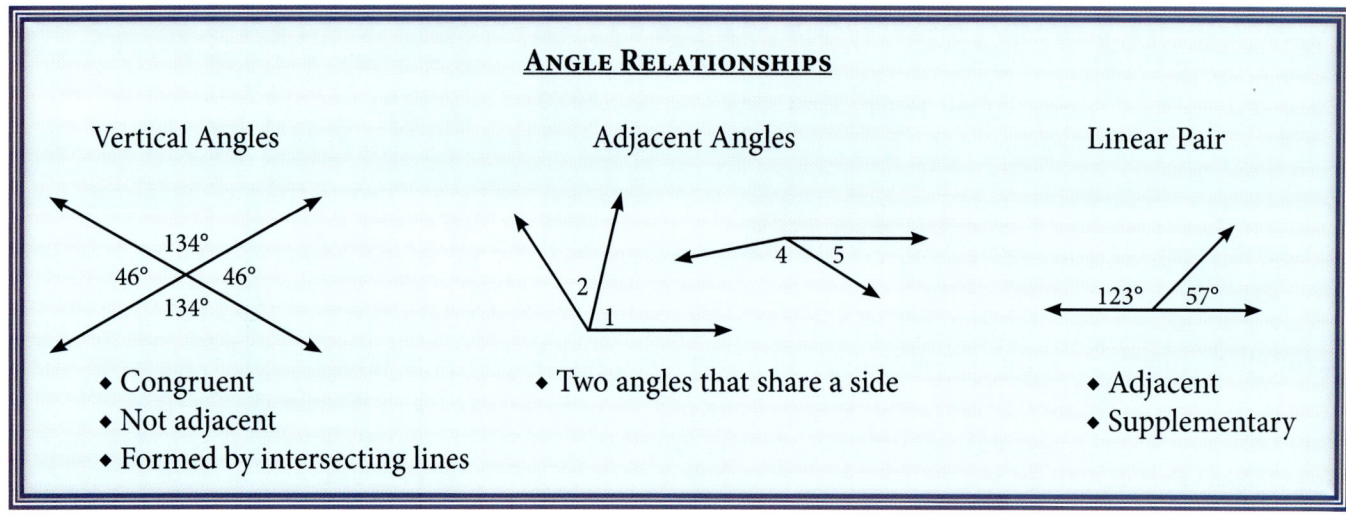

Sierra and King used different methods to find the solution to the question below. Look at Sierra's work and King's work.

Question	Sierra's Work	King's Work
If $m\angle 2 = 36°$ and $m\angle 3 = 144°$, what is the measure of $\angle 4$? 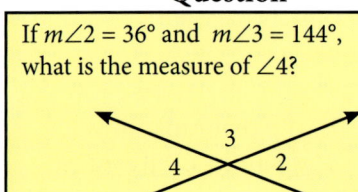	Sierra knows that $\angle 4$ and $\angle 3$ are supplementary because they are a linear pair. She subtracted 144° from 180°. $180° - 144° = 36°$ $m\angle 4 = 36°$	King knows that $\angle 2$ and $\angle 4$ are vertical angles. They have the same degree measure. $m\angle 4 = m\angle 2$ $m\angle 4 = 36°$

There is often more than one way to arrive at a correct answer. Both Sierra and King answered the question correctly but used different methods.

EXAMPLE 1 Find the measure of each missing angle.
  a. $m\angle 3$
  b. $m\angle 1$
  c. $m\angle 4$

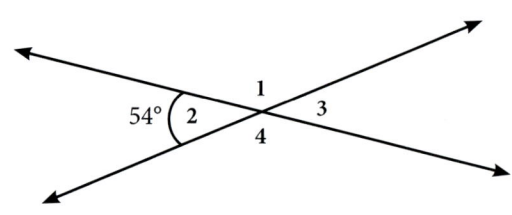

SOLUTIONS

a. Vertical angles are congruent. $m\angle 2 = m\angle 3$
$54° = m\angle 3$

b. $\angle 1$ and $\angle 2$ are a linear pair. $m\angle 1 + m\angle 2 = 180°$
Substitute 54° for $m\angle 2$. $m\angle 1 + 54° = 180°$
Subtract 54° from each side of the equation. $m\angle 1 = 126°$

c. Vertical angles are congruent. $m\angle 1 = m\angle 4$
$126° = m\angle 4$

EXAMPLE 2

Use the diagram at right.
a. Solve for *x*.
b. Find the measure of each angle.

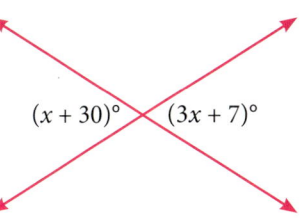

$(x + 30)°$ $(3x + 7)°$

SOLUTIONS

a. Vertical angles have equal measures.

$3x + 7 = x + 30$

Subtract *x* from each side of the equation.
Subtract 7 from each side of the equation.

$$2x + 7 = 30$$
$$-7 \quad -7$$
$$\frac{2x}{2} = \frac{23}{2}$$

Divide each side of the equation by 2.

$x = 11.5$

b. Write the given expression for each angle. $(3x + 7)°$ $(x + 30)°$
Substitute the solution for *x*. $(3(11.5) + 7)°$ $(11.5 + 30)°$
Multiply. $34.5 + 7$
Add. $41.5°$ $41.5°$

The measure of each angle is 41.5°.

Each special angle pair has properties that are important to remember. The sum of complementary angles is 90°. The sum of supplementary angles is 180°. Vertical angles are congruent. A linear pair has a sum of 180°.

EXERCISES

Use the diagram at right. Determine the measure of each unknown angle.

1. If $m\angle 1 = 50°$, find the following:
 a. $m\angle 2 = ?$
 b. $m\angle 3 = ?$
 c. $m\angle 4 = ?$

2. If $m\angle 4 = 153°$, find the following:
 a. $m\angle 1 = ?$
 b. $m\angle 2 = ?$
 c. $m\angle 3 = ?$

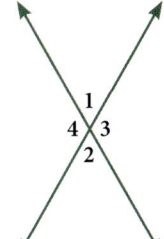

Sketch a diagram to represent each situation. Label each diagram.

3. A 30° angle and a 25° angle are adjacent.

4. $\angle ABC$ and $\angle CBD$ are complementary and adjacent.

5. $\angle 7$ and $\angle 8$ are vertical angles.

6. $\angle 9$ and $\angle 10$ are obtuse vertical angles.

7. $\angle 1$ and $\angle 2$ are a linear pair. $\angle 1$ is an acute angle.

8. $\angle JKL$ and $\angle LKM$ are adjacent and form a straight angle.

Lesson 1.4 ~ Vertical Angles and Adjacent Angles

Identify the angle relationship, then solve for *x*.

9.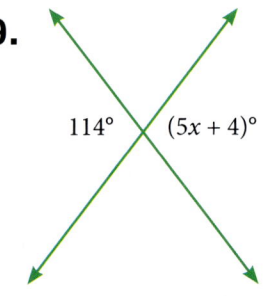

10.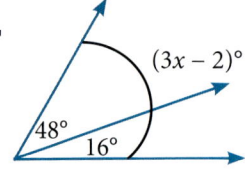

11.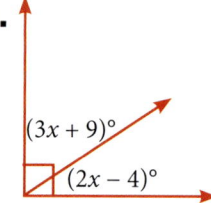

12.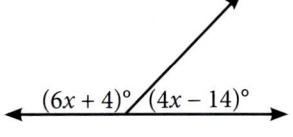

13.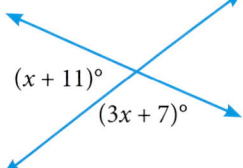

14.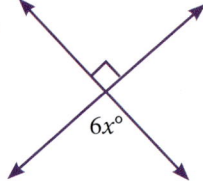

15.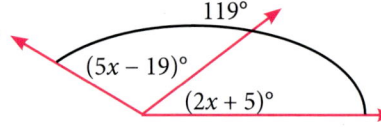

16.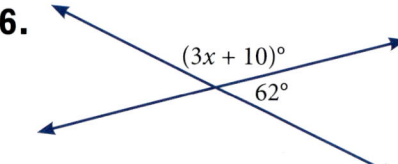

17.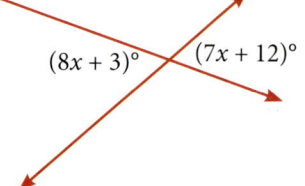

18. Use the diagram at the right.
    a. Find $m\angle ABC$.
    b. Find $m\angle CBD$.

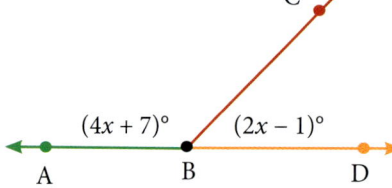

19. Jimmy found $m\angle TXY$ in the diagram below. His work is at right. There is an error in Jimmy's reasoning. Explain Jimmy's error and then find $m\angle TXY$.

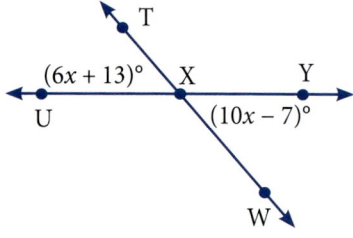

Jimmy's Work

$m\angle WXY = m\angle TXU$
$10x - 7 = 6x + 13$
$20 = 4x$
$5 = x$

$m\angle WXY = 10(5) - 7 = 43°$

$m\angle WXY + m\angle TXY = 90°$
$43° + m\angle TXY = 90°$

$\boxed{m\angle TXY = 47°}$

Determine which statement is true. Explain your reasoning.

20. Adjacent angles are linear pairs. OR Linear pairs are adjacent.

21. Vertical angles are adjacent. OR Vertical angles are formed by intersecting lines.

22 *Lesson 1.4 ~ Vertical Angles and Adjacent Angles*

REVIEW

Match the given diagram or information to a description from the word bank. Some diagrams or information may match more than one description.

Acute angle	Obtuse angle	Right angle
Congruent angles	Straight angle	Vertical angles
Supplementary angles	Linear pair	Complementary angles

22.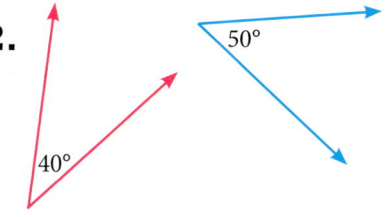

23.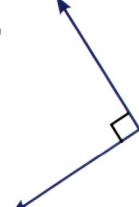

24.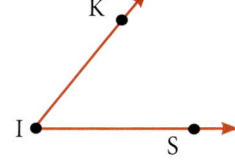

25. $m\angle PQR = 110°$
$m\angle ABC = 70°$

26.

27.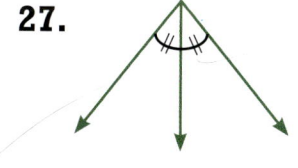

28.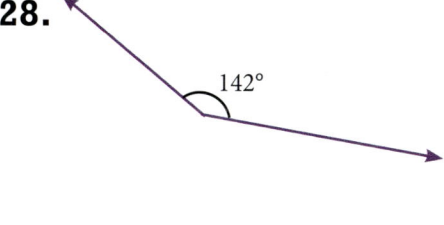

29.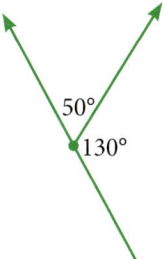

30.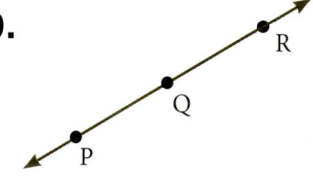

TIC-TAC-TOE ~ A TRANSVERSAL OF ANGLES

Step 1: Use the internet or math textbooks to research what a transversal line is. Draw a diagram showing a transversal and write a definition using your own words.

Step 2: Special angle pairs are formed by two parallel lines cut by a transversal. Name at least 4 different special angle pairs that are formed by two parallel lines and a transversal.

Step 3: Create a flip book with the special angle pairs you researched and the ones from this block. Include a diagram and definition for each special angle pair.

Lesson 1.4 ~ Vertical Angles and Adjacent Angles

DRAWING GEOMETRIC SHAPES

LESSON 1.5

 Draw geometric shapes with given conditions.

What makes a triangle? Three angles and three sides form a triangle. Are there any limitations when trying to draw a triangle? Will any three lengths work? Will any three angles form a triangle? If two triangles have the same angle measures, are they congruent? If the two triangles have the same side lengths, are they congruent? You will discover the answers to these questions in the following two-part **Explore!**.

EXPLORE!   **KNOWING THREE MEASURES**

Part 1: Three Angles

Step 1: Use a protractor and straightedge to draw a triangle with angles that are 90°, 20° and 70°.

Step 2: Write the angle measures inside each corresponding vertex. Cut out the triangle.

Step 3: Tape your triangle in the place your teacher has designated for this group of triangles.

Step 4: Next, use a protractor and straightedge to construct a triangle with two angles that are 125° and 20°. Determine the measure of the third angle. Write the angle measure inside each vertex of the triangle.

Step 5: Cut the triangle out and tape it in the place your teacher has designated for this set of triangles.

Step 6: Look at the first group of triangles. What do you observe about the size and shape of these triangles?

Step 7: Look at the second group of triangles. What do you notice about the angle measures, shape and size of each triangle?

Step 8: When you compare two triangles with three congruent angles, will the triangles always be unique? In other words, is there only one possible triangle that can be formed by given congruent angles? Be specific in terms of size and shape.

In *Part 1* of the **Explore!** you found that when three angles in one triangle are congruent to three angles in another triangle the two triangles may not be congruent. The two triangles have the same shape but may not be the same size. When two triangles have congruent angles and proportional sides they are called **similar**. It is also true that when two angles are congruent to two angles in another triangle the triangles are similar. When this occurs, the third angles in each triangle will also be congruent to each other.

ANGLE - ANGLE SIMILARITY RULE

When two angles in one triangle are congruent to two angles in another triangle, the triangles will be similar.

EXPLORE! **KNOWING THREE MEASURES**

Part 2: Three Sides

Step 1: Cut out thin strips of paper in the following lengths: 4 *cm*, 6 *cm*, 7 *cm*, 10 *cm*, 12 *cm*, 13.5 *cm*, 15 *cm*, 24 *cm*.

Step 2: Copy tables like the ones below on a piece of paper.

Lengths that Form a Triangle

Short Side	Medium Side	Long Side

Lengths that DO NOT Form a Triangle

Short Side	Medium Side	Long Side

Step 3: Use the following sets of three lengths of strips of paper to try to form a triangle. Record the lengths in the appropriate table.

    4 *cm*, 6 *cm*, 12 *cm*   7 *cm*, 10 *cm*, 15 *cm*   13.5 *cm*, 15 *cm*, 24 *cm*   7 *cm*, 10 *cm*, 24 *cm*

Step 4: Use the strips of paper to find two additional sets of strips of paper that form a triangle. Find two additional sets that do not form a triangle. Record each set of lengths in the appropriate table.

Step 5: In the table labeled "Lengths that Form a Triangle", calculate the sum of the short and medium sides of each triangle. What do you notice about the sum of the two shortest sides compared to the length of the longest side? Explain.

Step 6: Repeat **Step 5** for the table labeled "Lengths that DO NOT Form a Triangle". Explain what you notice about the sum of the lengths of the two shortest sides compared to the length of the longest side in this table.

Step 7: Using your conclusions in **Step 5** and **Step 6**, determine if the following lengths will form a triangle.
    a. 5 *in*, 7 *in*, 14 *in*   **b.** 12 *in*, 15 *in*, 23 *in*   **c.** 8 *ft*, 8 *ft*, 8 *ft*   **d.** 12.4 *m*, 15 *m*, 27.2 *m*

Step 8: Determine if lengths of 8 *cm*, 10 *cm* and 18 *cm* can form a triangle. Explain what you did to reach your conclusion.

Step 9: Write a rule about the lengths of the three sides of a triangle.

SIDE LENGTH INEQUALITY RULE

The sum of the lengths of any two sides of a triangle must be greater than the third side.

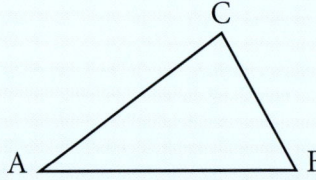

AB + AC > BC
AC + CB > AB
AB + BC > AC

When determining if three lengths will form a triangle it is only necessary to verify that the sum of the lengths of the two shortest sides is greater than the length of the longest side.

EXAMPLE 1 Determine if each set of side lengths can form a triangle.
a. 4 cm, 9 cm, 15 cm
b. 2 in, 2 in, 4 in
c. 14 m, 9.4 m, 8.3 m

SOLUTIONS

Write each inequality comparing the sum of the two shortest sides to the longest.

a. Is 4 + 9 > 15?   b. Is 2 + 2 > 4?   c. Is 8.3 + 9.4 > 14?

Simplify.   13 > 15   4 > 4   17.7 > 14

Is the sum more than the longest side?   No   No   Yes

a. No, the lengths 4 cm, 9 cm and 15 cm cannot form a triangle.
b. No, the lengths 2 in, 2 in and 4 in cannot form a triangle.
c. Yes, the lengths 14 m, 9.4 m and 8.3 m can form a triangle.

EXAMPLE 2 A triangle has two sides that are 7 feet and 12 feet. Find the values the third side must be between.

SOLUTION

Write each inequality using the known side lengths.
Let x be the unknown side.

If x is the short or medium side: $7 + x > 12$
$x > 5$

If x is the longest side: $12 + 7 > x$
$19 > x$

The third side must be greater than 5, but less than 19.

Rachel and Jackson each drew a triangle with matching side lengths. When they finished they noticed that their triangles looked exactly the same. Both of them made another triangle with the same side lengths and all the triangles were congruent to one another. Their triangles are represented below.

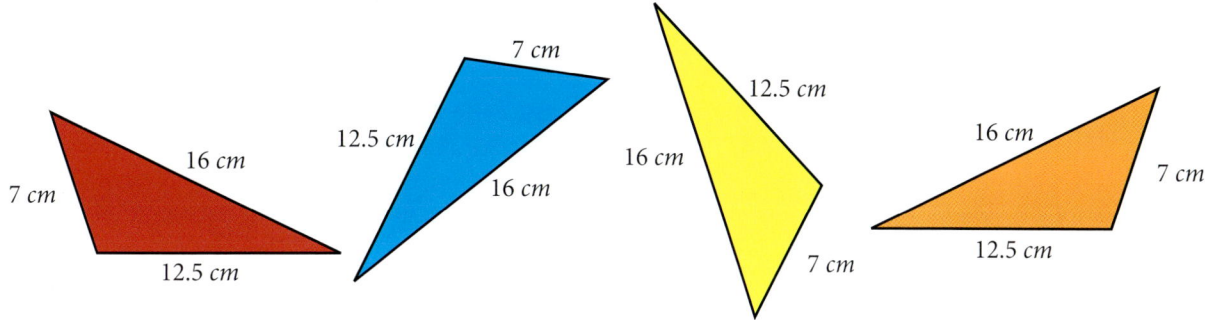

Triangles are congruent when three sides in one triangle are congruent to the corresponding sides in another triangle.

SIDE-SIDE-SIDE TRIANGLE CONGRUENCE

If all three sides in one triangle are the same length as the corresponding sides in another triangle, then the triangles are congruent.

EXERCISES

Determine if each statement is *ALWAYS*, *SOMETIMES* or *NEVER* true. Explain your reasoning.

1. The sum of the two shortest sides of a triangle must be less than the longest side.

2. If two angles of one triangle are congruent to two angles of another triangle, then the triangles are similar.

3. If three sides of one triangle are congruent to three sides of another triangle, the triangles are congruent.

4. In every pair of congruent triangles, there are 3 sets of corresponding congruent sides and 3 sets of corresponding congruent angles.

5. If two figures are similar, then they are congruent.

6. If two angles in one triangle are congruent to two angles in another triangle, the two triangles are congruent.

Determine if each set of lengths will form a triangle.

7. 5 cm, 9 cm, 12 cm

8. 3.6 m, 7.2 m, 10.9 m

9. 14 ft, 23 ft, 19 ft

10. 8.5 cm, 8.5 cm, 8.5 cm

11. 38 in, 21 in, 13 in

12. 1.5 km, 1.1 km, 0.4 km

Determine if each pair of triangles is similar. Explain your reasoning.

13.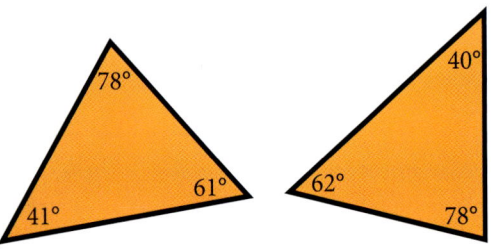

14.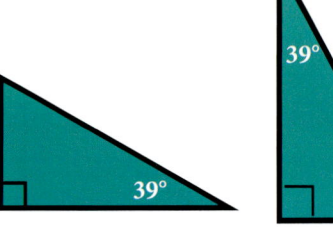

15. Construct two triangles with sides that are 5 *cm*, 3 *cm* and 7 *cm*. Label the first one △CAT and the second one △DOG. Are they congruent? Explain your reasoning.

16. A triangle has two sides that are 3 feet and 11 feet long. Which values below could be the length of the third side? Write all that apply.

   5 feet   8 feet   9 feet   10.7 feet   14 feet   16.5 feet

Two sides of a triangle are given. Determine the lengths the third side must be between.

17. 12 inches, 17 inches   **18.** 2.5 *cm*, 4.7 *cm*   **19.** 42 *ft*, 32.5 *ft*

20. Manuel knows the longest side length of a triangle is 8 inches. The other side lengths are integer measures. List all possible combinations of the other two side lengths.

For Exercises 21–26, use the triangles below.

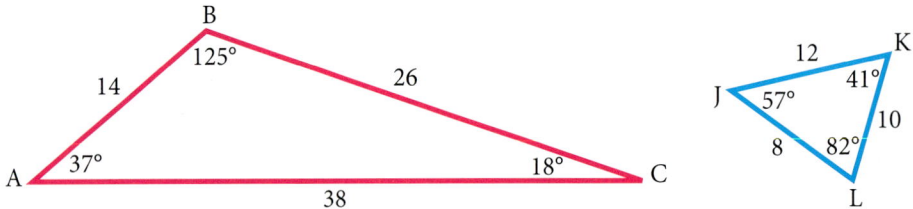

21. Is △ABC congruent to △JKL? Explain your reasoning.

22. Is △ABC similar to △JKL? Explain your reasoning.

23. Sketch a triangle that is similar to △ABC but not congruent. Label each side and each angle with appropriate measures.

24. Sketch a triangle that is similar to △JKL but is not congruent. Label each side and each angle with appropriate measures.

25. Sketch △VWT making it congruent to △ABC. Label each side and each angle with appropriate measures.

26. Sketch △MOP making it congruent to △JKL. Label each side and each angle with appropriate measures.

27. Prior to the start of a sailboat race, the judges must make sure all of the sails are the same shape. Explain how the judges can verify that each sail is the same shape without taking the sails down.

28   *Lesson 1.5 ~ Drawing Geometric Shapes*

REVIEW

Write two possible names for each angle.

28.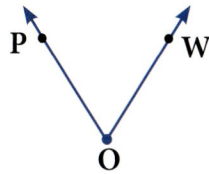

29.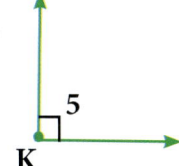

30.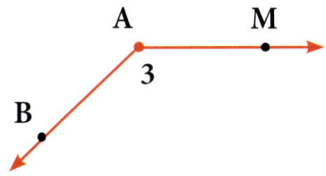

Sketch and label a diagram for each description.

31. an acute angle

32. vertical angles with each angle measuring 40°

33. adjacent supplementary angles

34. complementary angles that are not adjacent

Tic-Tac-Toe ~ Puzzling Angles

Find the measure of the numbered angles in the diagram. You may need to gather additional information about angles before completing the puzzle.
   a. Find what the sum of the angles in a triangle is.
   b. Find what the angles of a quadrilateral add up to.

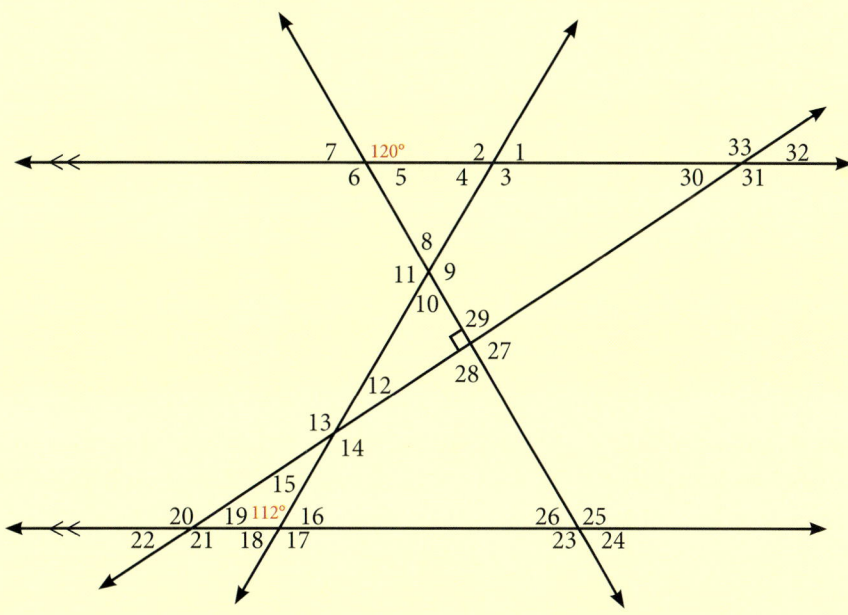

Tic-Tac-Toe ~ Crossword

Create a crossword puzzle using all of the vocabulary from **Block 1**. Make a blank master copy and an answer key.

REVIEW

BLOCK 1

Vocabulary

acute angle
adjacent angles
angle
complementary angles
congruent

degree
linear pair
obtuse angle
protractor
ray
right angle

similar
straight angle
supplementary angles
vertex
vertical angles

Measure, name and draw angles.
Classify angles as acute, right, obtuse or straight.
Use facts about complementary and supplementary angles to solve problems.
Use facts about vertical and adjacent angles to solve problems.
Draw geometric shapes with given conditions.

Lesson 1.1 ~ Measuring and Naming Angles

Use a protractor to measure each angle to the nearest degree.

1.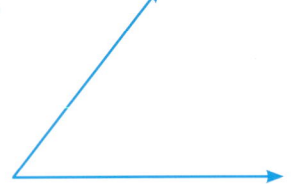

2.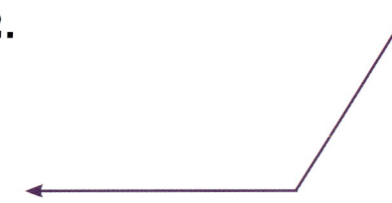

3.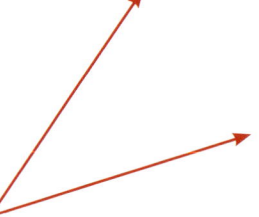

List the four names for each angle.

4.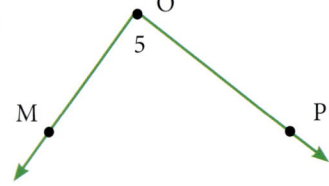

5.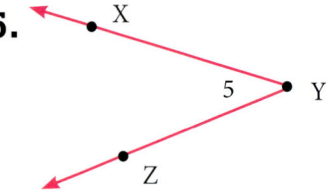

Sketch and label a diagram to represent each angle. Use a protractor, as needed.

6. $m\angle YOU = 125°$

7. $m\angle BAT = 40°$

8. $\angle HAM$ and $\angle HAP$ are adjacent

Lesson 1.2 ~ Classifying Angles

Classify each angle as acute, right, obtuse or straight.

9.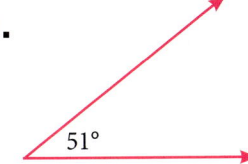

10.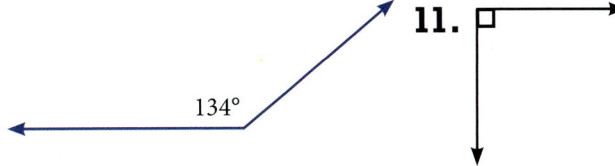

11.

Sketch and label a diagram to represent each statement.

12. ∠CAT ≅ ∠DOG

13. a right angle that can be named three different ways

Write an equation and solve for x. Check each solution.

14.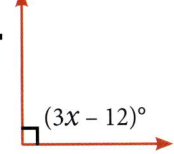

15.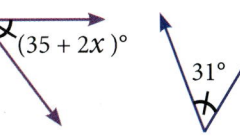

16.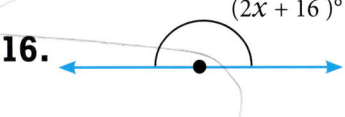

17. $m\angle BIG = (x + 14)°$
   a. What must x be equal to if ∠BIG is a right angle?
   b. What values must x be between if ∠BIG is acute?
   c. Jaxen says x must be greater than 76 and less than 180 if ∠BIG is obtuse. Is he correct? Explain your reasoning.

Lesson 1.3 ~ Complementary and Supplementary Angles

Identify each pair of angles as complementary, supplementary or neither.

18.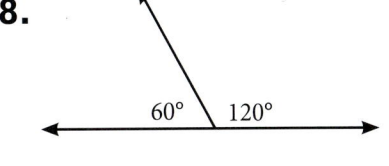

19.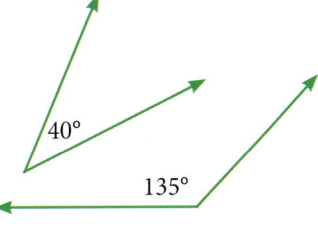

20.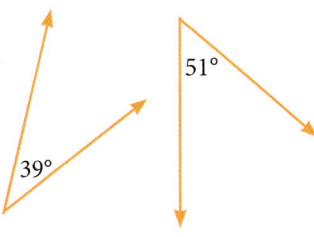

Solve for x. Check each solution.

21.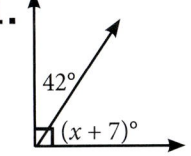

22. $m\angle A = (25 + 2x)°$
    $m\angle P = (10 + 3x)°$
    ∠A and ∠P are complementary.

23.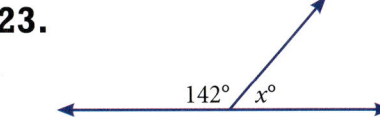

Block 1 ~ Review   31

Sketch and label a diagram for each situation. Solve for x.

24. ∠PAR and ∠TYE are supplementary; m∠PAR = 83° and m∠TYE = (x + 5)°.

25. ∠F and ∠G are supplementary; m∠F = 46° and m∠G = (3x − 25)°.

26. ∠1 and ∠2 are complementary; m∠1 = 2x° and m∠2 = 3x°.

Lesson 1.4 ~ Vertical Angles and Adjacent Angles

Determine the measure of the angles labeled a, b and c.

27.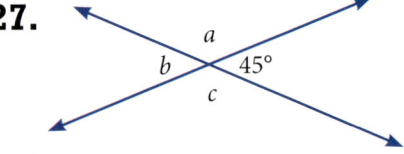

28.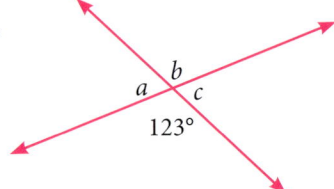

Name the special angle pair. Solve for x. Check your solution.

29.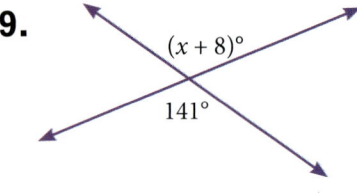

30.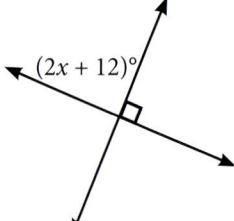

31.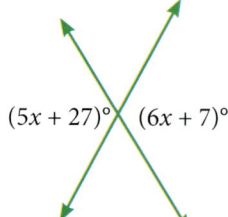

32.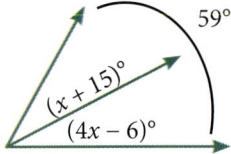

33.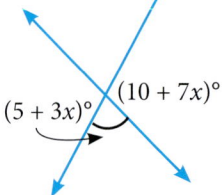

34.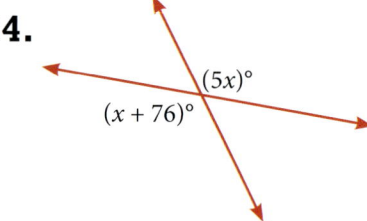

Lesson 1.5 ~ Drawing Geometric Shapes

Determine if each set of lengths will form a triangle.

35. 5 cm, 7 cm, 12 cm

36. 8 ft, 9.5 ft, 9.5 ft

37. 12.5 m, 5.9 m, 7.2 m

Two side lengths of a triangle are given. Determine what lengths the third side of each triangle must be between.

38. 8 inches, 17 inches

39. 2.3 cm, 6.8 cm

Determine if each pair of triangles are similar. Explain your reasoning.

40. 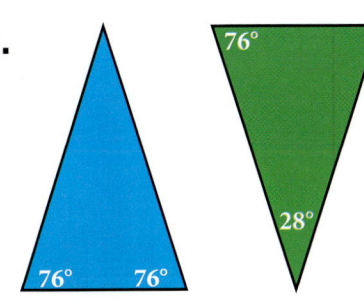   **41.**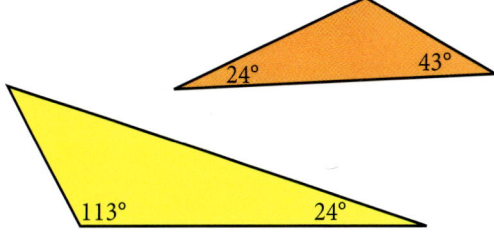

Determine if each statement is true or false. If the statement is false rewrite it to make it true.

42. The sum of the lengths of the two shortest sides of a triangle must be less than the length of the longest side.

43. If two angles in one triangle are congruent to two angles in another triangle then the triangles are congruent to one another.

44. If three sides of one triangle are congruent to three sides of another triangle, the triangles will have corresponding angles that are congruent.

Tic-Tac-Toe ~ Hidden Triangles

 **Step 1:** Find the total number of triangles in each figure.

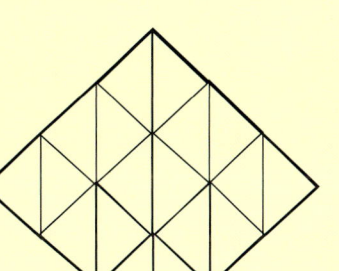

Step 2: Create your own triangle design.
Step 3: Count the number of triangles in your design.
Step 4: Make four copies of the triangle design and give it to four peers. Have them count the number of triangles in the design. Did they count the same number of triangles as you? What triangles did they find that you did not?

Tic-Tac-Toe ~ Triangle Collage

 **Step 1:** Find at least 2 photographs or pictures of each type of triangle (acute, right, obtuse, scalene, isosceles and equilateral). You can take your own photographs, locate and print pictures from the internet or cut pictures from magazines or newspapers.
Step 2: Identify and name each triangle. Then, provide a definition for each.
Step 3: Make a collage to display the pictures with their definition.

CAREER FOCUS

JERRY
NURSERY MANAGER

I am a nursery manager. I work with a staff planning and planting 25 different tree species. Some of the seedlings I plant are very recognizable. I plant Douglas Fir, Noble Fir, Ponderosa Pine and Giant Sequoia. Some of the trees I grow are for planting around homes. Others go into the woods for reforestation. When loggers cut down trees, the trees I grow are planted in their place. This helps make sure that there will be trees to harvest in the future. I help grow 6.5 to 7.5 million trees for about 150 different customers. I work at a small nursery, but some nurseries in the Northwest grow over 25 million trees a year.

I use math every day in my job. When preparing to plant, I have to know how many trees will fit on each acre. Some seedlings are planted at 75 per foot and some are planted at 18 per foot. Math helps me to determine how many seeds I will need to make sure I will have enough trees to fill a customer's order. Fertilizers are another area where I use math. I take soil samples and calculate at what rate I should apply fertilizer to make sure that the seedlings grow well. When I prepare to ship trees to a customer, I also have to use math to make sure that they are counted correctly and packaged right for each customer.

Most nursery managers have a Bachelor of Science degree in Agriculture or a related horticultural field. These degrees require high-level math classes like calculus. About 10% of nursery managers do not have a college degree. Those managers usually work their way up to a manager position with many years of work and experience. Nursery manager salaries start at around $35,000 to $40,000 per year and can get as high as $65,000 or more per year. Often the pay is related to the size of the nursery and how many trees the nursery sells.

I feel lucky to be doing something I enjoy and making a positive contribution to the world I live in by growing trees. I have helped grow over 180,000,000 trees in my career. I enjoy going to the woods and seeing stands of trees I grew. I know that some of those trees will still be living when my grandchildren's grandchildren are old.

CORE FOCUS ON SHAPES & ANGLES
BLOCK 2 ~ TWO-DIMENSIONAL GEOMETRY

LESSON 2.1	AREAS OF TRIANGLES AND PARALLELOGRAMS	37
LESSON 2.2	AREA OF A TRAPEZOID	42
	EXPLORE! A FORMULA FOR TRAPEZOID AREA	
LESSON 2.3	PARTS OF A CIRCLE	47
LESSON 2.4	CIRCUMFERENCE AND PI	52
	EXPLORE! A SPECIAL RATIO	
LESSON 2.5	AREA OF A CIRCLE	56
	EXPLORE! CIRCLE AREAS	
LESSON 2.6	MORE PI	61
	EXPLORE! WHICH PI?	
LESSON 2.7	COMPOSITE FIGURES	65
LESSON 2.8	CIRCLE SIMILARITY	69
	EXPLORE! STEPPING STONES	
LESSON 2.9	AREA OF SECTORS	74
REVIEW	BLOCK 2 ~ TWO-DIMENSIONAL GEOMETRY	78

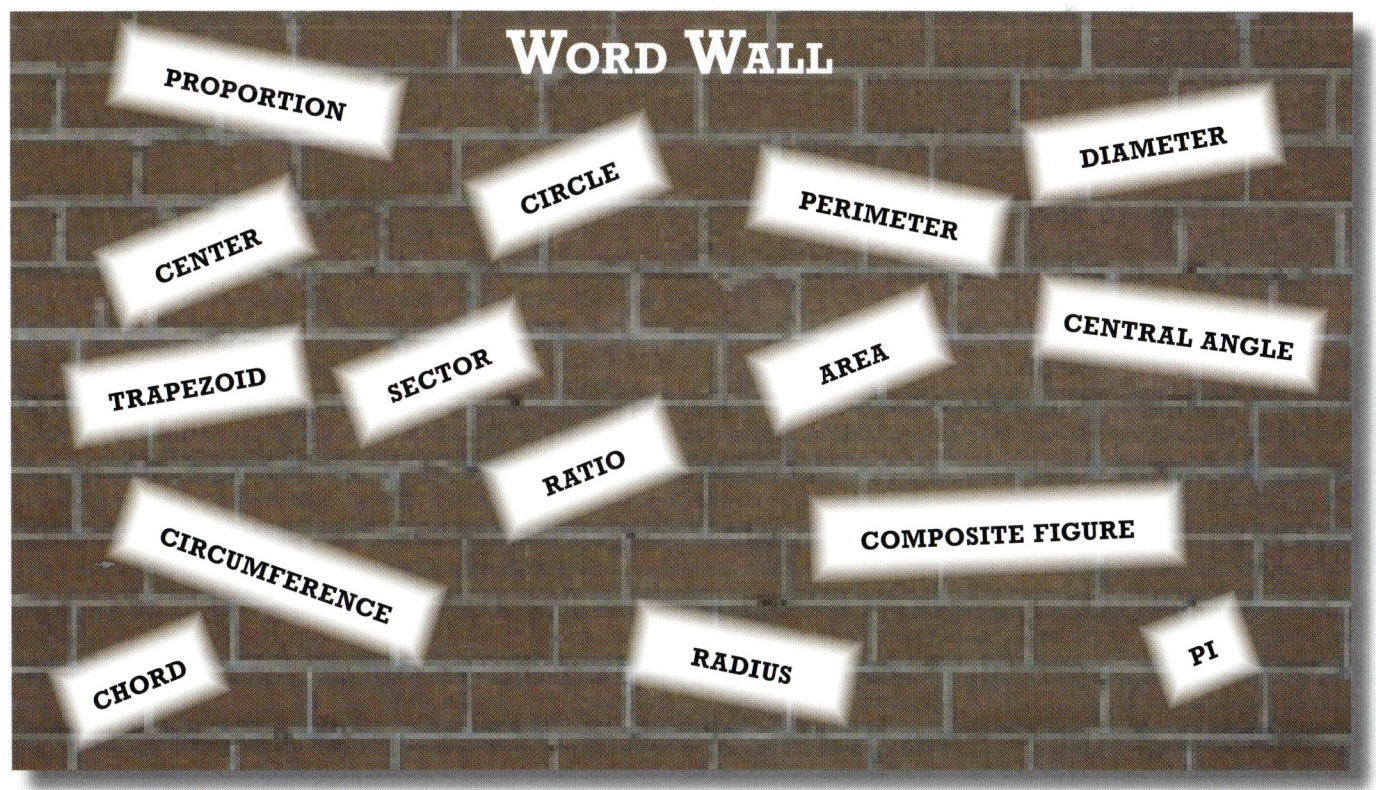

BLOCK 2 ~ TWO-DIMENSIONAL GEOMETRY
TIC-TAC-TOE

Find Your Foot

Estimate the area of your foot using grid paper. Find a combination of shapes that make your foot.

See page 41 for details.

More Quadrilateral Area

Investigate area formulas for kites and rhombi.

See page 51 for details.

Ferris Wheels

Research the first Ferris wheel. Write a paper.

See page 60 for details.

Flashcards

Create flashcards for use with this block.

See page 46 for details.

Equal Areas

Find the measures of individual geometric shapes that all have the same area.

See page 64 for details.

A Song About Pi

Demonstrate your understanding of pi by writing a song about the number and its uses.

See page 55 for details.

Pie Charts

Find the areas of sectors on a pie chart.

See page 77 for details.

Greeting Cards

Create a package of greeting cards using geometric shapes.

See page 64 for details.

Circles and Squares

Find how many circles will be in the next shape. Calculate the shaded region and find another pattern.

See page 73 for details.

AREAS OF TRIANGLES AND PARALLELOGRAMS

LESSON 2.1

 Use area formulas to compute area of geometric shapes and find missing measures.

Area is the number of square units needed to cover a space. Look at the rectangle below. The base is 4 units. The height is 2 units. There are 8 square units covering the space inside the rectangle. The area of the rectangle is 8 square units.

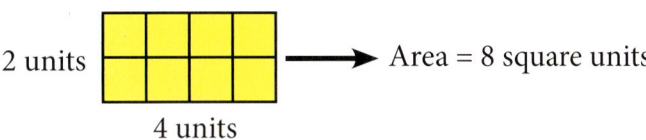

In previous years, you have learned how to find areas of four basic geometric shapes. The area of a shape is found by substituting information about a shape into the appropriate area formula.

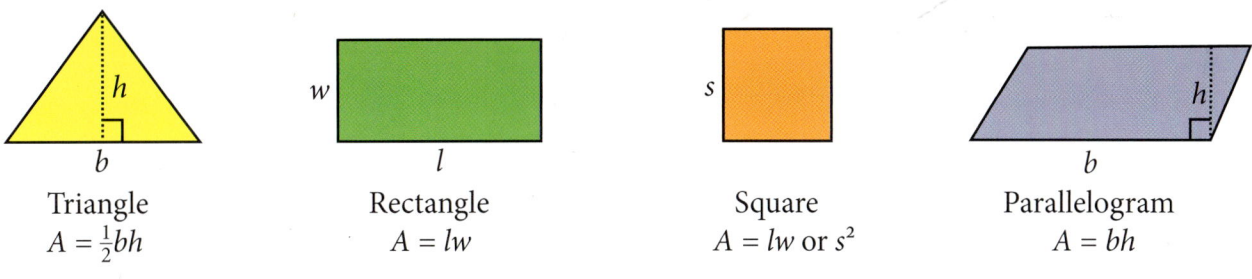

Triangle
$A = \frac{1}{2}bh$

Rectangle
$A = lw$

Square
$A = lw$ or s^2

Parallelogram
$A = bh$

EXAMPLE 1 Find the area of the triangle.

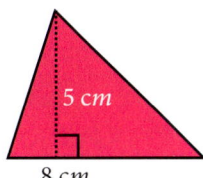

SOLUTION

Write the formula.   Area $= \frac{1}{2}bh$
Substitute the given measures. Area $= \frac{1}{2}(8)(5)$
Multiply.   Area $= 20$

The area of the triangle is 20 square centimeters.

> Square centimeters can also be written cm^2.

You can also work backwards to find a missing measure on a figure using a formula. Substitute all known values into the formula. Then solve the equation for the unknown measure.

Lesson 2.1 ~ Areas of Triangles and Parallelograms **37**

> **FINDING A MISSING MEASURE**
> 1. Find the formula which fits the given figure.
> 2. Substitute all known values for the variables in the formula.
> 3. Determine if any numbers on the same side of the equals sign can be combined. If so, combine those numbers.
> 4. Write the answer in a complete sentence.

EXAMPLE 2

Find each missing measure.

a. $A = 72\ m^2$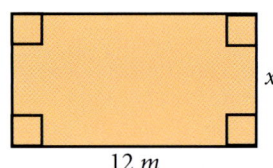

b. $A = 35\ in^2$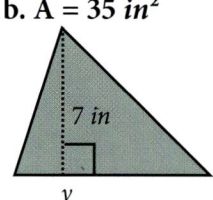

SOLUTIONS

a. Write the formula for the area of a rectangle.   Area = lw

Substitute all known values. $72 = 12x$

Divide both sides of the equation by 12. $\dfrac{72}{12} = \dfrac{\cancel{12}x}{\cancel{12}}$

$6 = x$

The width of the rectangle is 6 meters.

b. Write the formula for the area of a triangle.   Area = $\tfrac{1}{2}bh$

Substitute all known values. $35 = \tfrac{1}{2}(y)(7)$

Multiply the numbers on the right hand side of the equation ($\tfrac{1}{2}$ and 7). $35 = 3.5y$

Divide both sides of the equation by 3.5. $\dfrac{35}{3.5} = \dfrac{\cancel{3.5}y}{\cancel{3.5}}$

$y = 10$

The base of the triangle is 10 inches.

Perimeter is the distance around the outside of a figure. There are times when you need to use your knowledge of perimeter to find the area of a figure.

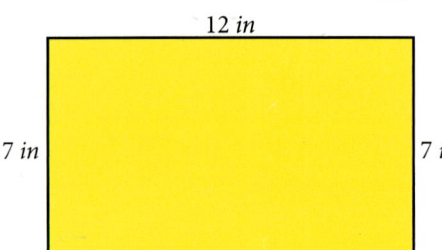

Add all of the side lengths to find the perimeter of the rectangle.
7 in + 12 in + 7 in + 12 in = 38 in
The perimeter of the rectangle is 38 inches.

Lesson 2.1 ~ Areas of Triangles and Parallelograms

EXAMPLE 3   The perimeter of a square bathroom tile is 32 centimeters. Find the area of the tile.

SOLUTION   Each side of a square is the same length.
Divide the perimeter by 4 to find the length of one side.
Write the formula for the area of a square.
Substitute the value for the variable.

$32 \div 4 = 8$
Area $= s^2$
Area $= 8^2$
Area $= 64$

The area of the bathroom tile is 64 square centimeters.

EXERCISES

Find the area of each figure.

1.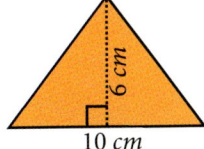

2.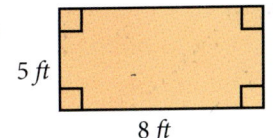

3.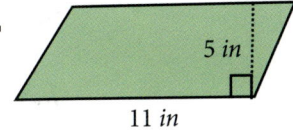

4.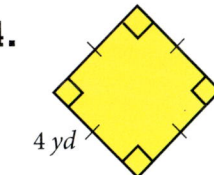

5.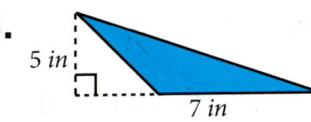

6.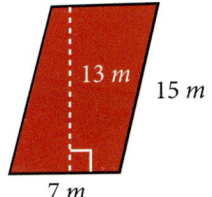

Sketch a diagram of each figure and label it with the given information. Calculate the area of each figure.

7. A triangle has a base of 8 *cm* and a height of 5 *cm*.

8. The base of a parallelogram is 21 feet and the height is 9 feet.

9. The width of a rectangle is 4.5 inches. The length is 3.5 inches longer than the width.

Find the missing measure.

10. A = 20 *in²*
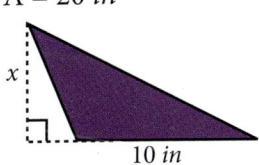

11. A = 42 *cm²*
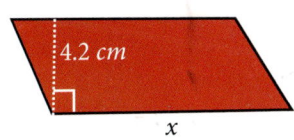

12. A = 45 square units
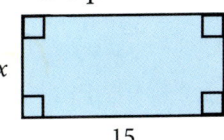

Find the missing measure.

13. A = 81 m²
9 m
x

14. A = 52 ft²
20 ft
x

15. A = 19.5 m²
x
14.2 m
3.9 m

16. The perimeter of a square is 72 m. Find the area of the square. Use mathematics to justify your answer.

17. The length of a rectangle is 2.5 cm. The area is 20 cm². What is the width of the rectangle? Marta's work to answer the question is at right. Identify her error and determine the width of the rectangle.

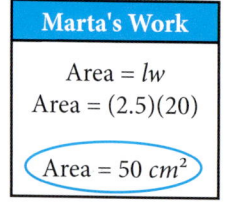

Marta's Work
Area = lw
Area = (2.5)(20)
Area = 50 cm²

18. A triangle has a height of 11 ft and an area of 110 ft². What is the length of the base?

19. Lois built a rectangular flower bed. She used four boards for the border of the flower bed. Two of the boards were 4 feet long. The other boards were 8 feet long.
  a. What was the area of the flower bed using these boards?
  b. She decided to cut each board in half to make two square flower beds. What was the area of each square flower bed?
  c. Did one or two flower beds give Lois more total area to plant flowers? Explain your reasoning.

20. A football field is 120 yards long and 160 feet wide. What is the area of a football field in square feet? Show all work necessary to justify your answer.

21. A window has 8 panes. Each pane measures 8 inches by 10 inches. What is the area of the entire window?

Plot each set of points on a coordinate plane. Connect the points in the order given. Find the area of each figure.

22. (3, 5), (3, 1) and (9, 1)

23. (0, −4), (0, 0), (−4, 0) and (−4, −4)

24. (−2, 5), (−2, −1), (4, −1) and (4, 5)

25. (0, 3), (5, 3), (4, −1) and (−1, −1)

26. Find the perimeter of each figure in **Exercises 23 and 24**.

27. Estimate the perimeter of the figure in **Exercise 25**. Are you able to find the exact perimeter? Why or why not?

28. Misty received a formula sheet from her math teacher. She noticed that the formula for the area of a square was not on it. Misty told her teacher it was missing. Her teacher says that the area of a square can be found by using the rectangle area formula. Do you agree? Explain your reasoning.

REVIEW

Use a protractor to measure each angle.

29.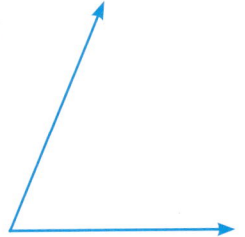

30.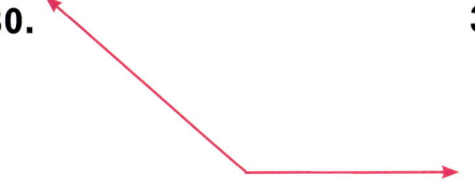

31.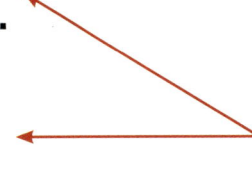

Draw an angle that matches each description. Use a protractor to measure the angle, then write the measure in the angle.

32. An acute angle

33. An obtuse angle

34. A right angle

Solve for *x*.

35. ∠A and ∠B are supplementary.

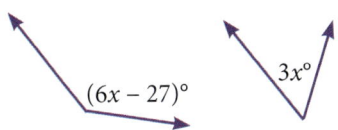

36.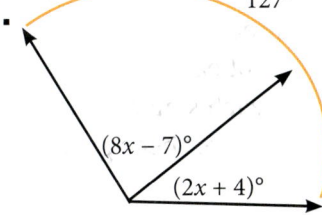

37.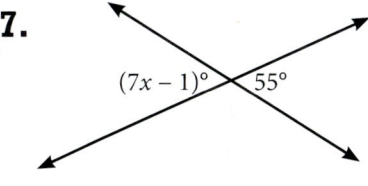

TIC-TAC-TOE ~ FIND YOUR FOOT

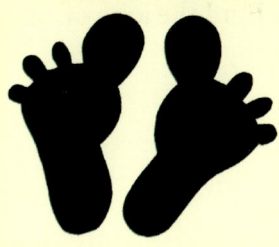

Formulas are used for common geometric figures. What happens if something is not a common geometric figure? Estimation is important to give an idea of a figure's area. You will estimate the area of your foot in this activity.

Step 1: Trace your foot on a piece of grid paper.

Step 2: Estimate how many square units your foot is. Each square on the paper represents one square unit.

Step 3: Look at the shape of your foot. Piece together geometric figures (squares, triangles, parallelograms and rectangles) that cover your foot outline.

Step 4: Calculate the area of each figure on a piece of paper. Find the sum of these areas.

Step 5: Compare your estimated answer with the sum of the areas of the geometric figures.

Step 6: Write two to three paragraphs discussing your observations, conclusions and thoughts on which method for calculating the area of your foot was best.

AREA OF A TRAPEZOID

LESSON 2.2

Understand and use the trapezoid area formula.

A **trapezoid** is a quadrilateral with two parallel sides. The figures below are trapezoids.

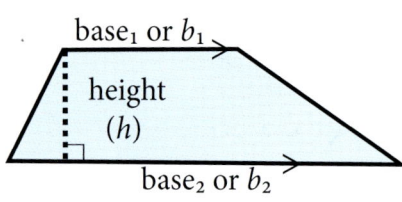

The parallel sides of a trapezoid are called the bases. A trapezoid has two bases. Each is identified using a subscript number. Subscript numbers identify objects that have the same name but represent different parts. The height of a trapezoid is the shortest distance between the bases which is a segment perpendicular to both bases.

EXPLORE!   A FORMULA FOR TRAPEZOID AREA

Step 1: Trace the trapezoid shown. Label b_1, b_2 and h.

Step 2: Two trapezoids can be connected to form a parallelogram. Trace the trapezoid a second time so it connects with the first one, forming a parallelogram. *(Hint: you may have to turn your paper.)* Label b_1, b_2 and h on the second trapezoid.

Step 3: Write an expression for the base of the parallelogram.

Step 4: Write an expression for the area of the parallelogram.

Step 5: How does the area of one trapezoid compare to the area of the parallelogram? Rewrite the expression from **Step 4** to represent the area of one of the trapezoids.

Step 6: Use the formula you developed in **Step 5** to calculate the area of each trapezoid below.

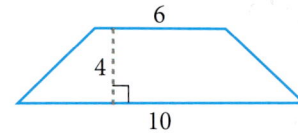 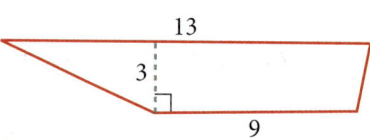

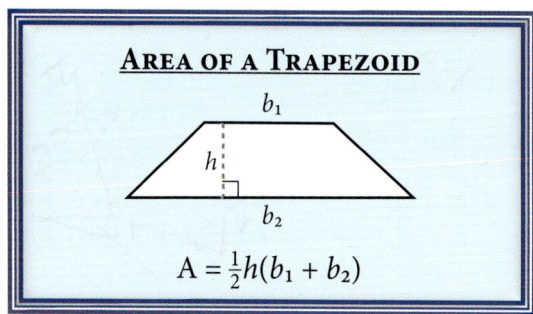

AREA OF A TRAPEZOID

$$A = \tfrac{1}{2}h(b_1 + b_2)$$

42   Lesson 2.2 ~ Area of a Trapezoid

EXAMPLE 1 Find the area of the trapezoid.

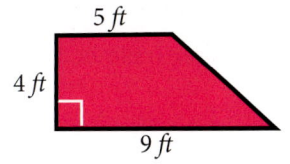

SOLUTION

Write the trapezoid area formula.   Area = $\frac{1}{2}h(b_1 + b_2)$

Substitute all known values.   Area = $\frac{1}{2}(4)(5 + 9)$

Add the numbers in parentheses.   Area = $\frac{1}{2}(4)(14)$

Multiply.   Area = $2(14) = 28$

The area of the trapezoid is 28 ft^2.

A missing measure on a trapezoid can be found by substituting the known values into the formula. Once all the known values have been substituted, you can solve the equation for the missing measure.

EXAMPLE 2 Find the height of the trapezoid. The area of the trapezoid is 17.5 cm^2.

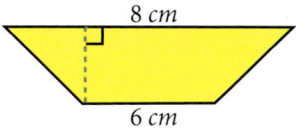

SOLUTION

Write the trapezoid area formula.   Area = $\frac{1}{2}h(b_1 + b_2)$

Substitute all known values. $17.5 = \frac{1}{2}h(8 + 6)$

Add the numbers in parentheses. $17.5 = \frac{1}{2}h(14)$

Multiply $\frac{1}{2}$ and 14. $17.5 = 7h$

Divide by 7 on both sides of the equation. $\dfrac{17.5}{7} = \dfrac{\cancel{7}h}{\cancel{7}}$

                                            $2.5 = h$

The height of the trapezoid is 2.5 *cm*.

Lesson 2.2 ~ Area of a Trapezoid   **43**

EXAMPLE 3

Nakisha painted a large trapezoid-shaped mural. The area of the mural is 88 square feet. The longer base is 20 feet. The height is 6.4 feet. Find the length of the missing base.

SOLUTION

Draw a figure to represent the given information.

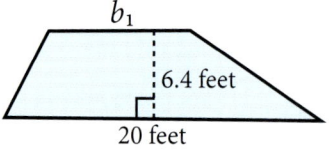

Write the trapezoid area formula.   Area = $\frac{1}{2}h(b_1 + b_2)$

Substitute all known values. $88 = \frac{1}{2}(6.4)(b_1 + 20)$

Multiply $\frac{1}{2}$ and 6.4. $88 = 3.2(b_1 + 20)$

Distribute. $88 = 3.2(b_1 + 20)$
$88 = 3.2b_1 + 64$

Subtract 64 from both sides of the equation. $\dfrac{-64 \quad\quad -64}{\dfrac{24}{3.2} = \dfrac{3.2b_1}{3.2}}$

Divide by 3.2.

$7.5 = b_1$

The length of the missing base on the mural is 7.5 feet.

EXERCISES

Calculate the area of each trapezoid.

1.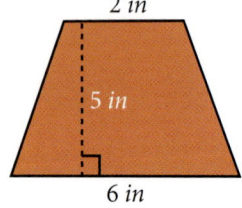

2. Height = 3 m
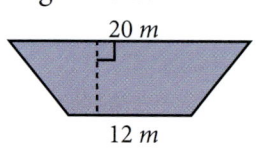

3.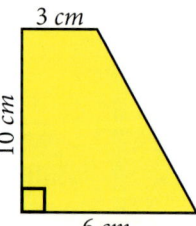

4.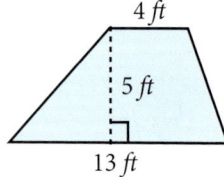

5.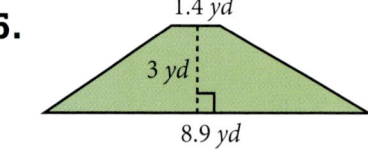

6.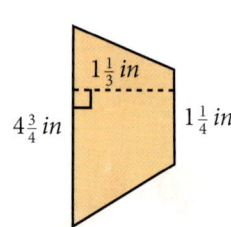

7. Find the area of a trapezoid with the following measures: b_1 = 24 inches, b_2 = 30 inches, h = 6 inches.

8. A trapezoid has a height of 12.3 miles. The bases measure 3.5 miles and 14.5 miles. Find the area.

9. The bases of a trapezoid measure 6 feet and 14 feet. The height of the trapezoid is 7 feet. Find the area.

10. Celeste created a patio that used bricks with a trapezoidal shape on the top and bottom of each brick. Each trapezoid had base measures of 20 cm and 35 cm. The distance between the bases (height) was 12 cm.
  a. What is the area of the top trapezoid on one brick?
  b. It took 812 bricks to cover the patio. How many square centimeters did the patio cover?

Find the area of each figure.

11.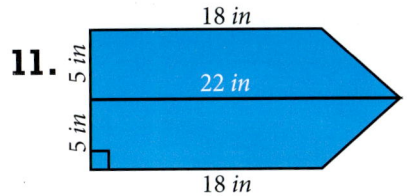

12.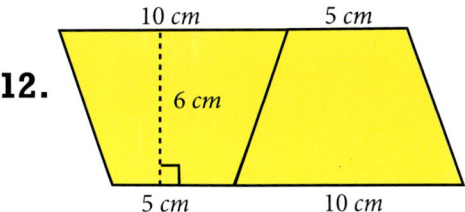

Find the unknown base or height of each trapezoid.

13. A = 60 ft²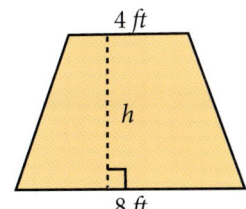

14. A = 77 in²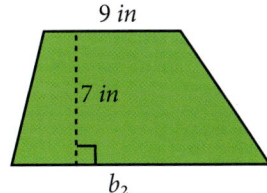

15. A = 27.03 m²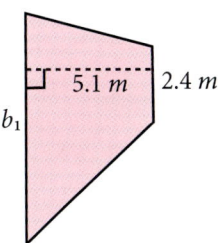

16. A trapezoid has an area of 62 square inches. The bases measure 8 and 12 inches. Find the height.

17. Find the missing base measurement for a trapezoid with the following measures:
$b_1 = 9$ meters, $h = 6$ meters and $A = 51\ m^2$

Manny's Work

$A = \frac{1}{2}h(b_1 + b_2)$
$102 = \frac{1}{2}(12)(8 + b_2)$
$102 = 6(8 + b_2)$
$102 = 48 + b_2$
$54 = b_2$

The base is 54 ft.

18. The area of a trapezoid is 102 square feet. The height is 12 feet. One of the bases is 8 feet long. Manny's work to find the missing base is at left. Identify his mistake and find the length of the missing base.

19. The area of a trapezoid is 100 square inches. The height is 10 inches. Give two pairs of possible lengths for the bases.

20. One base of a trapezoidal photo frame is 10 inches. The other base measures 18 inches. The area of the frame is 224 square inches. Find the height of the frame.

21. Find the sum of the triangle areas.
  a. Find the area of the top triangle (A).
  b. Find the area of the bottom triangle (B).
  c. Find the sum of the triangles' area.
  d. Use the trapezoid area formula to calculate the area. Does your answer match **part c**? Explain why or why not.

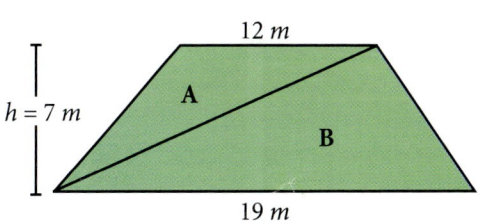

Lesson 2.2 ~ Area of a Trapezoid

22. Use the blue trapezoid at right.
   a. Find the area of the triangle on the left.
   b. Find the area of the rectangle in the middle.
   c. Find the area of the triangle on the right.
   d. Add the answers from **a–c** to find the area of the trapezoid.
   e. Find the length of b_1 and b_2.
   f. Calculate the area of the trapezoid using the trapezoid area formula.
   g. Compare and contrast the two different methods used to find the area of the trapezoid. Describe a time when finding the area of each piece may be more helpful than using the formula.

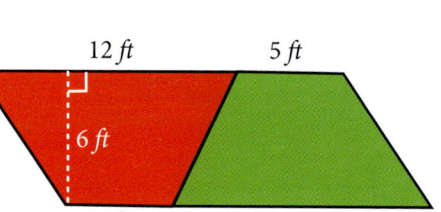

23. Use the figure to the right.
   a. Find the area of the red and green trapezoids.
   b. Add the areas together.
   c. What shape is the entire figure?
   d. Calculate the area of the entire figure using the formula $A = bh$.
   e. Are the answers in **parts b and d** the same? Explain your reasoning.

REVIEW

Find the area of each figure.

24.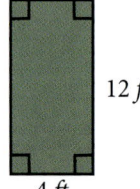

25.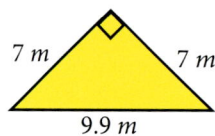

26.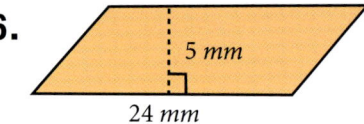

Find the missing measure.

27. $A = 90\ in^2$
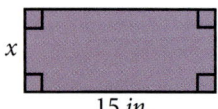

28. $A = 25$ square units
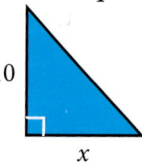

29. $A = 33\ m^2$
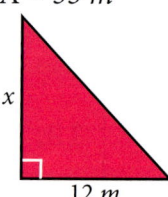

TIC-TAC-TOE ~ FLASHCARDS

Create a set of 30 flashcards to help a student review the material in **Block 2**. Use cardstock paper or blank index cards. Write the question on one side of the card and the answer on the other side. The set should include at least three cards from each of the following categories:

- Areas of basic geometric figures
- Vocabulary
- Uses of the different estimates of π
- Areas of composite figures
- Word problems
- Parts of a circle

PARTS OF A CIRCLE

LESSON 2.3

 Identify, name and define parts of circles.

A **circle** is the set of all points that are the same distance from a point called the **center**. A circle is named using its center point. The circle to the right is named ⊙A, which is read "circle A."

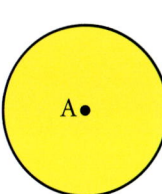

Vocabulary	Diagram	Definition
Radius		The distance from the center of a circle to any point on the circle.
Chord		A line segment with endpoints on the circle.
Diameter		The distance across a circle through the center. The diameter is the longest chord. It is made of 2 radii.
Central Angle		An angle with its vertex at the center of the circle.

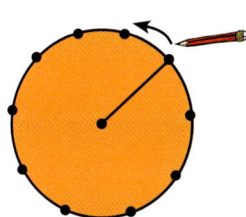 Circles can be drawn using a stencil, tracing around a circular object or using a compass. You can also sketch a circle by drawing points an equal distance from the center and connecting the points with a smooth curve.

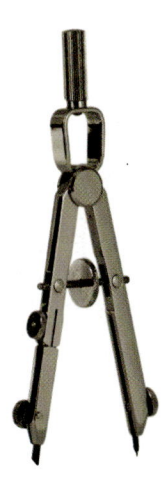

Lesson 2.3 ~ Parts of a Circle   47

In geometry, a line segment is named using its two endpoints.

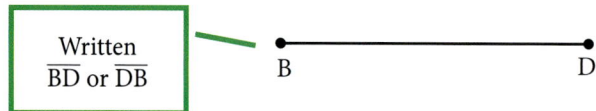

Diameters, radii, chords and central angles are named using points on the circle or inside the circle.

EXAMPLE 1   Use ⊙C to name each circle part.
a. A chord
b. A radius
c. A diameter
d. A central angle

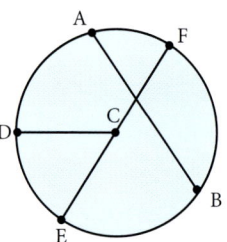

SOLUTIONS

a. Any segment with endpoints on the circle. $\overline{AB}$ or $\overline{EF}$
b. A segment from the center of the circle to $\overline{CD}$ or $\overline{CE}$ or $\overline{CF}$
   any point on the circle.
c. A chord that goes through the center. $\overline{EF}$
d. An angle with the vertex at the center of the circle.   ∠ECD, ∠FCD or ∠FCE

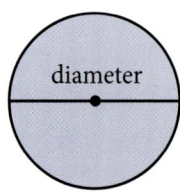

A diameter is a special chord. It has endpoints on the circle, but it also goes through the center of the circle. It is twice as long as a radius. The diameter is the distance across a circle through its center.
For a diameter d, and its radius, r: $d = 2r$ $r = \frac{1}{2}d$

EXAMPLE 2   a. The diameter of a circle is 12 centimeters. Find the radius.
b. The radius of a circle is 7 inches. Find the diameter.

SOLUTIONS

a. A radius is half the length of a diameter. $r = \frac{1}{2}d$

   Multiply the diameter by $\frac{1}{2}$. $r = \frac{1}{2}(12) = 6$

   The radius is 6 centimeters.

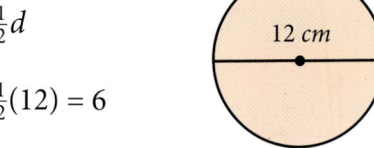

b. A diameter is twice as long as a radius. $d = 2r$

   Multiply the radius by 2. $d = 2(7) = 14$

   The diameter is 14 inches.

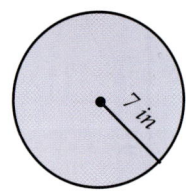

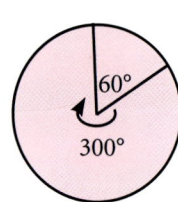

A central angle is an angle with its vertex at the center of the circle. It is measured using degrees. Each circle has a total of 360° around its center point. There are two central angles in the circle to the left.

Lesson 2.3 ~ Parts of a Circle

CENTRAL ANGLE SUM

The sum of the central angles in a circle is 360°.

EXAMPLE 3 Find the measure of the missing central angle.

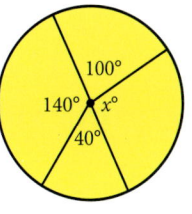

SOLUTION The sum of the four central angles is 360°.
Write an equation.
Combine like terms.
Subtract 280 from both sides of the equation.

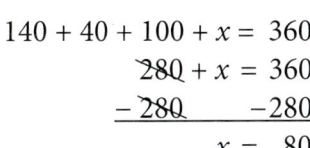

The measure of the missing angle is 80°.

EXERCISES

1. List at least 5 circular objects seen or used in everyday life.

2. What are three ways to draw a circle?

3. Draw ⊙K with radius $\overline{PK}$.

4. Draw a circle with chord $\overline{AB}$.

5. Draw a circle with radius $\overline{HM}$ and diameter $\overline{MX}$.

6. Draw ⊙R with central angle ∠NRE.

7. Draw a circle with a radius of 3 centimeters. Explain the process you used to draw the circle.

8. Draw a circle with a diameter of 2 inches.

Use ⊙B to name each of the following parts.

9. A radius

10. A diameter

11. A central angle

12. A chord

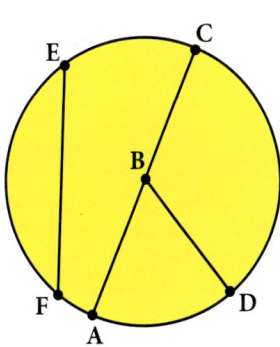

Lesson 2.3 ~ Parts of a Circle   49

In ⊙V, identify each of the following parts.

13. three radii

14. two diameters

15. two chords

16. three central angles

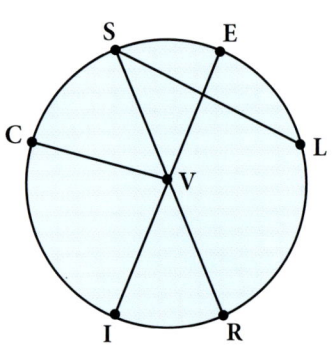

Find the diameter of a circle with each radius.

17. 28 m

18. 6.25 yd

19. $4\frac{2}{3}$ in

Find the radius of a circle with each diameter.

20. 8 cm

21. 15 yd

22. 9.4 mm

Solve for x.

23.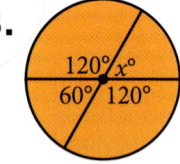

24.

25.

26.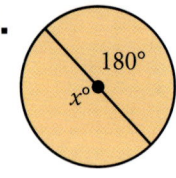

27.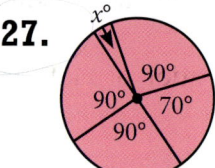

28.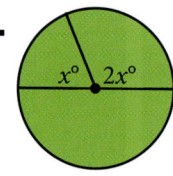

29. A circle has four central angles. Three of the angles measure 70° each. What is the measure of the fourth angle? Use mathematics to justify your answer.

30. ⊙W is divided in half. There are three central angles on one half of the circle. Two of the central angles measure 43° and 76°. What is the measure of the third central angle on that half of the circle? Show all work necessary to justify your answer.

31. To solve for x in the circle at right, Trevon used congruent vertical angles to determine x = 80. Penny noticed a linear pair and showed x + 100 = 180 which meant x = 80. Describe and show another way to find the value of x.

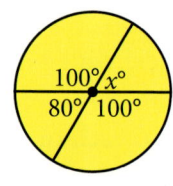

32. A circle is drawn with two radii that create two central angles. Which of the degree measurements below could be measures of those two central angles? Write all pairs of measures that apply.

| 80° and 100° | 300° and 60° | 180° and 180° |
| 90° and 270° | 200° and 100° | 40° and 320° |

50 Lesson 2.3 ~ Parts of a Circle

REVIEW

Calculate the area of each figure.

33.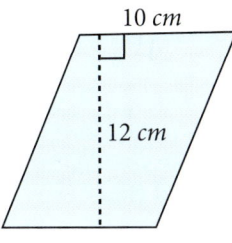

34.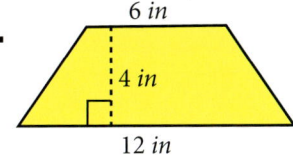

35.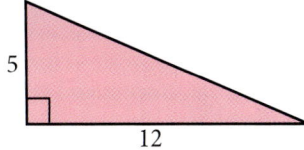

36.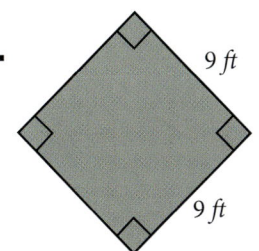

37.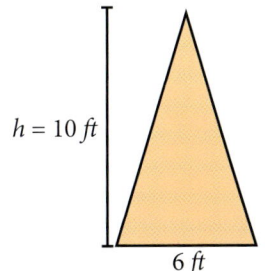

38.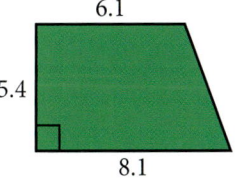

TIC-TAC-TOE ~ MORE QUADRILATERAL AREA

Quadrilaterals are geometric figures that have four sides. You have worked with squares, rectangles, trapezoids and parallelograms. Two additional types of quadrilaterals are the kite and the rhombus.

1. Find and record the definition of a kite and a rhombus. Draw an example of each.

2. List the special properties of a kite and a rhombus.

3. Find or develop the formulas used to calculate areas of a kite and a rhombus.

4. Copy each shape below onto your own paper. Find the area of each shape. Show all work.

A.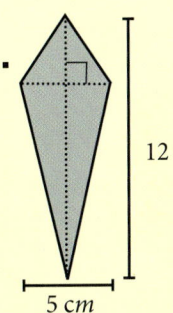

B.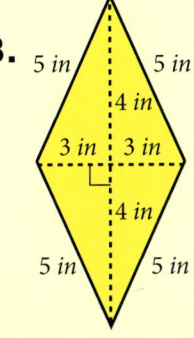

C.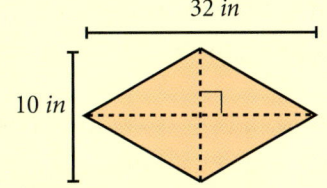

D.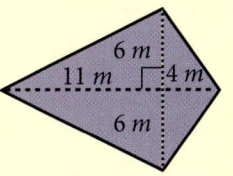

5. Draw two additional kites and two additional rhombi. Measure the key dimensions on each to the nearest tenth of a centimeter. Find the area of each figure.

Lesson 2.3 ~ Parts of a Circle **51**

CIRCUMFERENCE AND PI

LESSON 2.4

 Understand and use the relationship between pi and diameter to find circumference.

The distance around an object is called the perimeter. The distance around a circle is called the **circumference**. There is a special relationship between the circumference of a circle and its diameter.

EXPLORE!   A SPECIAL RATIO

Step 1: Gather 5 circular objects (such as lids, cups, cans, etc) to measure.

Step 2: Copy the chart.

Circular Object	Circumference	Diameter	Circumference / Diameter

Step 3: Write the name of each object in the column "Circular Object".

Step 4: Use a tape measure to find the circumference of each object to the nearest tenth of a centimeter. Record the measurement in the "Circumference" column in the chart.

Step 5: Use a tape measure or ruler to find the diameter of each object to the nearest tenth of a centimeter. Record the measurement in the "Diameter" column.

Step 6: Find the ratio of the circumference to the diameter for each object. Use a calculator. Write each answer in the last column as a decimal rounded to the nearest hundredth.

Step 7: What do you notice about the decimal values in the last column of the chart? Compare with a classmate.

Step 8: About how many times larger is the circumference than the diameter? Write a formula for finding the approximate circumference of a circle. $C \approx \_\_ \cdot d$

Step 9: Use the formula to find the approximate circumference of each circle.

a. 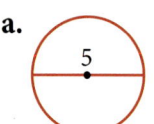   b. 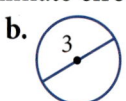   c.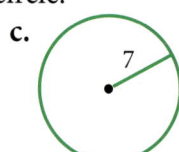

The circumference of a circle is a little bit more than three times the length of the diameter. You can wrap three diameters along the edge of a circle. There will be a small part of the circle not covered because the circumference is larger.

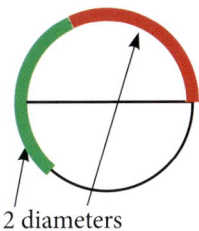
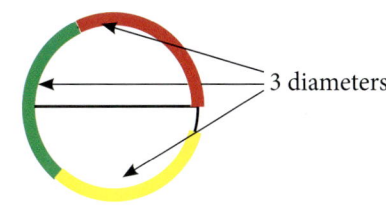

The exact number of times the diameter can be wrapped around the circle is represented by the Greek letter **π (pi)**. Pi is the ratio of the circumference of a circle to its diameter.

The exact value of π cannot be written as a decimal because it never terminates and never repeats. Most people round π to the nearest hundredth and use the number 3.14. Sometimes the fraction $\frac{22}{7}$ is used to estimate π.

> **CIRCUMFERENCE OF A CIRCLE**
>
> $C = \pi d = 2\pi r$
>
> The circumference of a circle is the product of π and the diameter of the circle.

EXAMPLE 1 Find the circumference of each circle. Use 3.14 for π.

a. (circle H, 8 mm)   b. (circle J, 2.5 ft)

C = Circumference
d = diameter
r = radius
$2r$ = diameter

SOLUTIONS

a. Write the circumference formula using the diameter. $C = \pi d$
Substitute the known values. $C \approx (3.14)(8)$
Multiply. $C \approx 25.12$
The circumference of ⊙H is about 25.12 mm.

b. Write the circumference formula using the radius. $C = 2\pi r$
Substitute the known values. $C \approx 2(3.14)(2.5)$
Multiply. $C \approx 15.7$
The circumference of ⊙J is about 15.7 feet.

EXAMPLE 2 Natalya walked with her friend around a circular track. The circumference of the track was 188.4 meters. Find the approximate length of the track's radius. Use 3.14 for π.

SOLUTION

Write the circumference formula. $C = 2\pi r$
Substitute the known values. $188.4 \approx 2(3.14)r$
Multiply. $188.4 \approx 6.28r$
Divide both sides by 6.28. $\frac{188.4}{6.28} \approx \frac{6.28r}{6.28}$
   $30 \approx r$

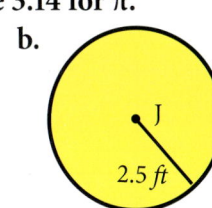

The radius of the track was approximately 30 meters.

EXERCISES

Find the circumference of each circle. Use 3.14 for π.

1.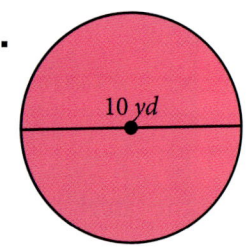
2.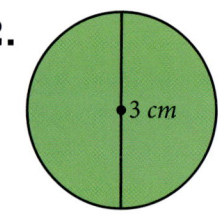
3.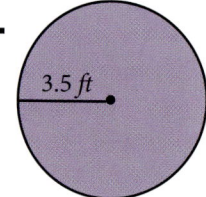

4.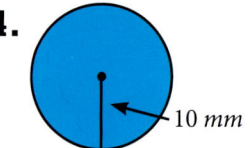
5.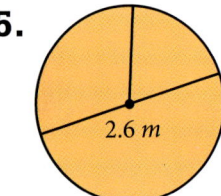
6.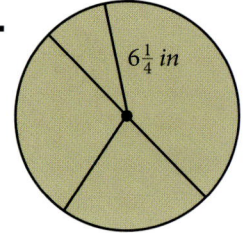

7. A van manufacturer recommends a wheel with a diameter of 17 inches.
   a. What is the circumference of this wheel?
   b. How far does the van travel each time the wheel makes one full rotation? Explain your reasoning.

8. A circular doughnut has a radius of 2 inches. Find the circumference.

9. A circular irrigation system is watering a field of alfalfa. The arm of the sprinkler is 1,250 feet long. Find the distance around the outside of the watered region.

10. The International Space Station orbits approximately 342 kilometers above the earth. The radius of Earth is 6,357 kilometers.
    a. Find the radius of the Space Station's orbit from the center of the earth.
    b. Find the distance the Space Station travels in one orbit.

Find each missing measure. Use 3.14 for π.

11. $C = 12.56\ ft$
    $d \approx ?$

12. $C = 94.2\ cm$
    $r \approx ?$

13. $C = 213.52\ in$
    $d \approx ?$

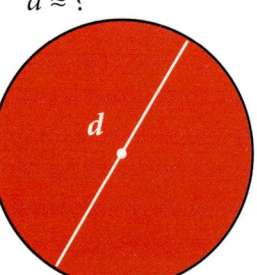

14. The circumference of a circle is 16.328 meters. Find the length of the diameter.

15. The circumference of a circle is 62.8 miles. What is the length of the radius?

Lesson 2.4 ~ Circumference and Pi

16. Find the length of the diameter of a circle whose circumference is 29.516 yards.

17. Julio paddled his boat around the edge of a circular lake. He traveled a total of 3.14 miles. What is the diameter of the lake?

18. The center circle on a basketball court has a circumference of 37.68 feet. Find the approximate diameter and radius of the circle at center court.

19. A tractor travels 169.5 inches on one rear tire revolution. What is the diameter of the tire? Round to the nearest whole number. Show all work necessary to justify your answer.

20. Which is greater: the perimeter of a square with side lengths of 10 *in* or the circumference of a circle with a diameter of 10 *in*? Use words and/or numbers to show how you determined your answer.

21. When finding the circumference of a circle with a diameter of 4 *ft*, Linda wrote C ≈ (3.14)(4). Why did she use the approximately equals to symbol (≈) in her work when the formula uses an equals to symbol, C = πd?

REVIEW

22. Sketch a circle with radius $\overline{AB}$ and a chord $\overline{AC}$.

23. Sketch a circle with diameter $\overline{WY}$ and a central angle named ∠YMP which measures 90°.

24. Sketch a triangle with a height of 1.5 inches and a base of 2 inches. Find the area of the triangle.

25. A triangle has an area of 42 square meters. The height of the triangle is 14 meters. What is the length of the base?

26. Sketch a trapezoid with bases of 3 and 5 centimeters and a height of 2 centimeters. Find the area of the trapezoid.

27. A trapezoid has an area of 94.5 square inches. The sum of the two bases is 21 inches. What is the height of the trapezoid?

Tic-Tac-Toe ~ A Song about Pi

Write a song about the number pi to a tune that is familiar to most people. Your song must have at least two verses and a chorus. Some ideas to include in your song are the different estimates of pi, the uses of pi, the formulas which use pi or the history of the number pi. Neatly write or type the lyrics of your song.

AREA OF A CIRCLE

LESSON 2.5

 Understand and use the circle area formula.

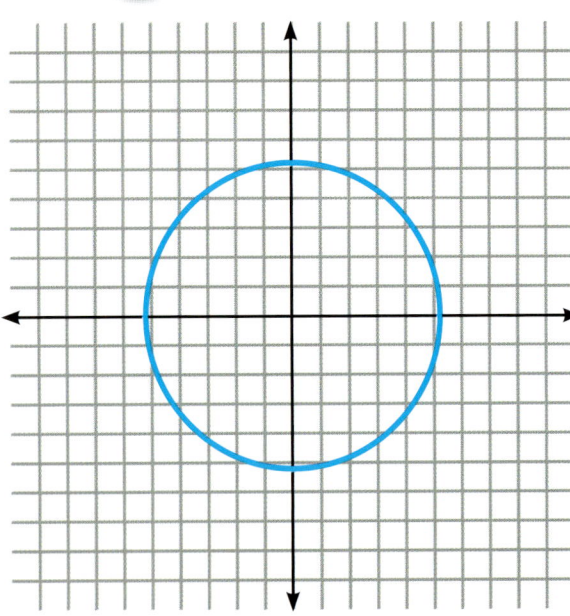

Maria and Greg are making circular coasters for a family dinner. They want the radius of each coaster to be 5 centimeters long. How many square centimeters will each coaster cover on a table?

Maria drew a coordinate plane. She sketched the circular coaster. She estimated the area by counting the number of squares inside the circle. She counted approximately 76 squares or parts of squares inside the drawing. Greg thought there might be a more accurate method to find the area of each coaster.

You will learn how to find the area of a circle using a formula in this lesson.

EXPLORE!   **CIRCLE AREAS**

Step 1: Trace a circle onto a piece of paper or use a compass to draw a circle. The radius of the circle should be at least 2 inches.

Step 2: Fold the circle in half three times to get 8 equal-sized parts.

Step 3: Cut the eight equal-sized parts of the circle on the fold lines.

Step 4: Arrange the pieces as shown.

Step 5: Which quadrilateral does the new figure resemble?

Step 6: Fill in the blanks.
    a. The height of the figure is equal to the _____ of the circle.
    b. The base of the figure is equal to half the _____ of the circle.

Step 7: How can the area of this figure be found? Create a formula to find the area of the circle.

Step 8: Use the formula to find the area of one coaster described at the beginning of this lesson.

The area formula of a circle can be determined by arranging pieces of a circle to form a shape similar to a parallelogram. The height of the parallelogram is the radius of the circle. The length of the base is half the circumference of the circle.

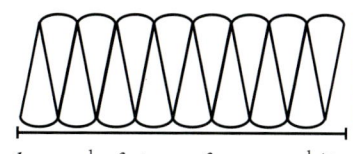

height = radius of circle

base = $\frac{1}{2}$ of circumference = $\frac{1}{2}(2\pi r)$

The area of a parallelogram is found by multiplying base times height. When the circle is arranged to form a parallelogram, the area of the circle can be determined by multiplying the base times the height.

$$Area = base \cdot height$$
$$Area = \tfrac{1}{2}C \cdot r$$
$$Area = \tfrac{1}{2}(2\pi r) \cdot r$$
$$Area = \pi r \cdot r$$
$$Area = \pi r^2$$

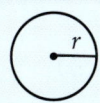

AREA OF A CIRCLE

$$A = \pi r^2$$

The area of a circle is the product of π and the square of the radius.

EXAMPLE 1   Find the area of the circle. Use 3.14 for π.

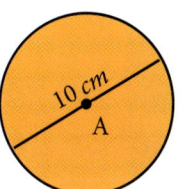

SOLUTION

Write the area formula.	$A = \pi r^2$
Find the length of the radius.	$r = 10 \div 2 = 5$
Substitute the known values.	$A \approx (3.14)(5)^2$
Square 5.	$A \approx (3.14)(25)$
Multiply.	$A \approx 78.5$

The diameter is given. However, the radius is needed. Divide the diameter by 2 to get the length of the radius.

The area of ⊙A is approximately 78.5 square centimeters.

The circle in **Example 1** is the same size as Maria's and Greg's coasters from the beginning of the Lesson. Each coaster will be approximately 78.5 square centimeters.

Lesson 2.5 ~ Area of a Circle

EXAMPLE 2

Jaira is having pizza for her birthday party. She can make two small pizzas, each with a radius of 6 inches, or one large pizza with a radius of 12 inches. Which option will give her more square inches of pizza?

SOLUTION

	Two small pizzas	One large pizza
Write the circle area formula.	$A = \pi r^2$	$A = \pi r^2$
Substitute the known values.	$A \approx (3.14)(6)^2$	$A \approx (3.14)(12)^2$
Square the radius.	$A \approx (3.14)(36)$	$A \approx (3.14)(144)$
Multiply.	$A \approx 113.04$	$A \approx 452.16$
Multiply the area of a small pizza by 2 for two small pizzas.	$A \approx 113.04(2) \approx 226.08$	

The large pizza will give Jaira twice as much pizza as two small pizzas.

Up to this point you have used a common estimate for pi, 3.14, for calculations. There will be situations when exact answers are needed for problems involving circles. For example, a manufacturer might program a machine to create a circular tarp with a radius of 3 feet. The exact area of one tarp needs to be calculated in order to find the amount of material used for one tarp. Estimating means the answer is not exact.

Exact answers are written using the π symbol. A common estimate of pi should not be substituted for pi when an exact answer is required.

EXAMPLE 3

Find the exact area of ⊙W.

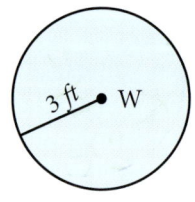

SOLUTION

Write the circle area formula.	$A = \pi r^2$
Substitute known values.	$A = \pi(3)^2$
Square 3.	$A = \pi(9)$
Rewrite with the number before the symbol.	$A = 9\pi$

The exact area of ⊙W is 9π square feet.

EXERCISES

Find the area of each circle. Use 3.14 for π.

1.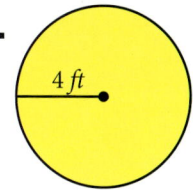

2.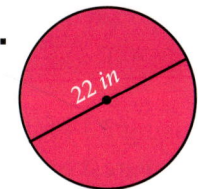

3.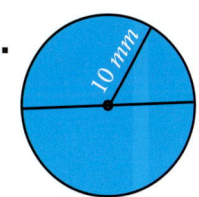

4. Mark set up a sprinkler to water the backyard. The sprinkler waters in a circular motion. The radius of the area covered by the sprinkler is 12 feet. Find the approximate square footage of the watered area.

Theo's Work
$A = \pi r^2$
$A \approx (3.14)(16)^2$
$A \approx (3.14)(32)$
$A \approx 100.48$
$\boxed{100.48 \ in^2}$

5. An extra large pizza has a radius of 16 inches. Theo miscalculated the area of the pizza in his work at left. Identify his error and determine the actual area of the pizza.

6. A circular tablecloth has a diameter of 14 feet. What is the area of the tablecloth?

7. A lighthouse beam reaches 21 miles in all directions. How many square miles does the light cover? Round the answer to the nearest square mile.

8. The diameter of a classroom clock face is 13 inches. Find the area of the clock face.

9. Lacey plans to put another window in her bedroom. She needs to decide whether to put in two small windows, each with a 1 foot radius, or one large window with a 2 foot radius. She wants as much sunlight as possible in her room. Which option should she use? Show all work necessary to justify your answer.

Calculate the exact area of each circle.

10.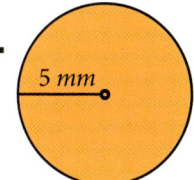

11.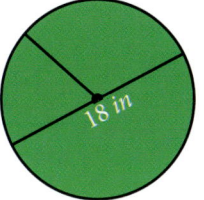

12. A circle with a diameter of 14.4 feet.

13. A milling machine is programmed to cut a hole with a radius of 8 millimeters. The computer program requires the programmer to enter the area of the hole. Find the exact area of the hole.

14. A circle has a radius of 1 inch. Find the exact area of the circle.

15. One circle has a diameter of 5 inches. Another circle has a radius of 2.5 inches. How do the areas of the circles compare? Use words and/or numbers to show how you determined your answer.

16. Sandra found the area of a circle to be exactly 16π square centimeters. If a circle has a larger area than Sandra's circle, what could the length of the radius be? Explain how you know your answer is correct.

17. A dart board has a circumference of 56.52 inches.
   a. Find the diameter of the dartboard. Use 3.14 for π.
   b. Find the approximate area of the dart board.

18. The exact circumference of a circle is 6π feet. Find the exact area of the circle. Show all work necessary to justify your answer.

19. The circumference of a circle is about 7.85 meters using 3.14 for π. Find the area of the circle. Round the answer to the nearest hundredth and use mathematics to justify your answer.

20. The area of a circle is $16\pi\ in^2$. What is the diameter?

21. Each circle has a radius of 4 *cm*. Find the exact area of each shaded region, given that the regions in each circle are equal in size.

a. 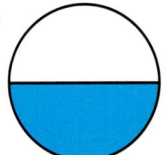   b. 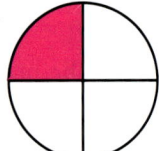   c. 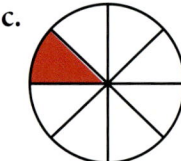   d.

REVIEW

Use the diagram at right.

22. Name the center.

23. Identify two radii.

24. Identify a central angle.

25. Name the shortest chord.

26. Name the longest chord.

27. What is the name of the circle?

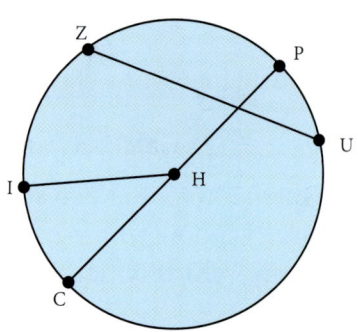

Tic-Tac-Toe ~ Ferris Wheels

Write a research paper about the Ferris wheel. It must be at least one page in length. Include the following:
- facts about the first Ferris wheel;
- information about 2 additional Ferris wheels;
- measurements for diameter, radius and circumference;
- a list of resources used to find the information;
- a diagram of one Ferris wheel you researched labeled with the measurements.

MORE PI

LESSON 2.6

 Identify and use common estimations of pi.

You have found exact answers for circumference and area using π. You have also used one common estimate of pi, 3.14. The fraction $\frac{22}{7}$ is another approximation of pi. Another common approximation of pi is the π button on a calculator.

EXPLORE! WHICH PI?

Step 1: Copy and complete the table below. Use each approximation of pi to find the circumference and area of ⊙A.

Approximation of Pi	Radius of Circle	Circumference (2πr)	Area (πr²)
3	7		
3.14	7		
$\frac{22}{7}$	7		
Calculator π	7		

Step 2: Most calculators show 3.141592654 as the value of π. Why is this considered an approximation?

Step 3: Which of the above approximations gives the least accurate answer? Explain your reasoning.

Step 4: Which of the above approximations gives the most accurate answer? Explain your reasoning.

Step 5: List the approximations of π from least to greatest.

Step 6: The most common estimate of π is 3.14. Why would this be the estimate used most often?

Step 7: For what type of numbers might $\frac{22}{7}$ be the most useful approximation to use for π?

Lesson 2.6 ~ More Pi **61**

EXAMPLE 1 Use the π button on your calculator to find the circumference and area of ⊙H. Round to the nearest hundredth.

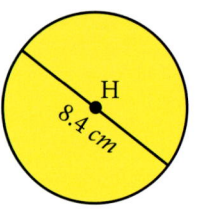

SOLUTION

	Circumference	Area
Write the formulas.	$C = \pi d$	$A = \pi r^2$
Substitute all known values.	$= (\pi)(8.4)$	$= \pi(4.2)^2$
Simplify.	$= 8.4\pi$	$= \pi(17.64)$
Multiply using π button.	≈ 26.38937829	≈ 55.41769441
Round to the nearest hundredth.	≈ 26.39	≈ 55.42

The radius is half the diameter.

The circumference of ⊙H is about 26.39 *cm* and the area of ⊙H is about 55.42 *cm²*.

EXAMPLE 2 Find the area and circumference of a circle with a 14 inch radius. Use $\frac{22}{7}$ for π.

SOLUTION

	Circumference	Area
Write the formulas.	$C = 2\pi r$	$A = \pi r^2$
Substitute all known values.	$\approx 2(\frac{22}{7})(14)$	$\approx (\frac{22}{7})(14)^2$
Rewrite with fractions.	$\approx (\frac{2}{1})(\frac{22}{7})(\frac{14}{1})$	$\approx (\frac{22}{7})(\frac{14}{1})(\frac{14}{1})$
Multiply.	$\approx \frac{616}{7}$	$\approx \frac{4312}{7}$
Divide.	$\approx 616 \div 7 = 88$	$\approx 4312 \div 7 = 616$

The circumference of the circle is about 88 inches. The circle area is about 616 *in²*.

CHOOSING COMMON ESTIMATES OF PI

3	Used as a very rough estimate of π and for mental calculations.
3.14	The most common estimate of π.
$\frac{22}{7}$	Best used when the radius is a multiple of 7 or if the radius is a fraction.
Calculator π	The most accurate estimate. Used when a calculator is available.

EXAMPLE 3 Choose the most appropriate estimate of pi for each description. Explain your reasoning for each choice.
a. Find the area of a circle with a radius of 21.
b. Find the circumference of a circle when you want a rough estimate.
c. Find the most accurate area of a circle using an approximation.

SOLUTIONS
a. The most appropriate estimate of π is $\frac{22}{7}$ since 21 is a multiple of 7.
b. Three can be used for π when only needing a rough estimate.
c. The π button on the calculator should be used when the most accurate estimate is needed.

EXERCISES

Identify the most appropriate estimate of pi to use in each situation. Explain your reasoning.

1. Habika orders bark dust for a circular flower bed. She is calculating the area of the flower bed.

2. Marlon programs a computer to make circular parts for dental equipment.

3. Hiroshi finds the circumference of a circle with a radius of 42 feet.

4. Gayle roughly estimates the circumference of her circular driveway.

 Use a calculator π button to find the circumference and area of each circle. Round to the nearest hundredth.

5.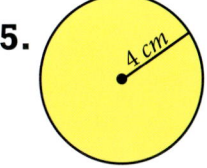

6.

7. ⊙P with a diameter of 12.5 cm

8. ⊙M with a radius of $2\frac{1}{2}$ ft

Find the circumference and area of each circle in Exercises 9–13. Use $\frac{22}{7}$ for π.

9. A circle with a radius of 14 feet.

10. A circle with a diameter of 70 meters.

11.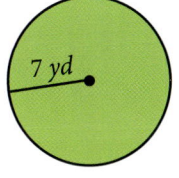

12.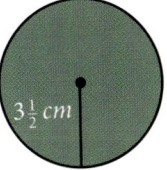

13.

Choose an appropriate estimate of pi, then answer each exercise.

14. Find a rough estimate of the area of a circular field with a radius of 110 feet.

15. Kekona knows the radius of her circular driveway is 20 feet. She wants to border it with decorative stone. Find a rough estimate of the circumference of Kekona's driveway.

16. Pizza comes in 12 inch or 14 inch diameters. Graysen really likes the crust. How much more crust will Graysen get if he orders a 14 inch pizza instead of a 12 inch pizza, to the nearest inch?

17. A twirled lasso has a radius of 6 feet. What is the area enclosed by the lasso's circle?

18. A circle has a radius of 35 mm. Shayla estimates the circumference of the circle using $\frac{22}{7}$. Tomiko uses 3.14 to estimate the circumference. By how much do their estimates differ? Use mathematics to justify your answer.

Lesson 2.6 ~ More Pi

19. ⊙M has a circumference of 14π centimeters. What is the area of ⊙M? Use mathematics to justify your thinking.

20. The first Ferris wheel had a diameter of 76 meters. How far did a person travel in 10 revolutions? Show all work necessary to justify your answer.

21. A circular pool has a diameter of 21 feet. The pool cover has an overhang of one foot. Find the circumference of the cover.

REVIEW

Write each area formula using variables. Identify the shape or shapes that each formula could be used for.

22. base times height

23. one-half base times height

24. radius squared times pi

25. one half the height times the quantity of base one plus base two

26. length times width

27. length of a side squared

Tic-Tac-Toe ~ Equal Areas

1. Draw a square that has an area of 36 square inches. Cut out the figure and label the lengths of the sides inside the square.

2. Create six more figures that all have areas of 36 square inches. The figures must include a triangle, a rectangle, a parallelogram, a trapezoid, a circle and a composite figure. Cut out each figure and label the key dimensions on each figure.

3. Make a poster displaying each geometric figure, its dimensions and its area calculation.

Tic-Tac-Toe ~ Greeting Cards

1. Design eight greeting cards. The set should include cards made of each of the following shapes:
   - Triangle
   - Square
   - Parallelogram
   - 2 Composite Figures
   - Rectangle
   - Trapezoid
   - Circle

2. Measure the key dimensions on each shape to the nearest tenth of a centimeter. Find the area of each cut-out and record it along with the dimensions on the back of the greeting card.

3. Illustrate or color each card. Write a different greeting or expression on each card.

4. Package the cards creatively as if they will be displayed and sold at a store.

COMPOSITE FIGURES

LESSON 2.7

Find the area of composite figures.

Composite figures are made up of two or more geometric shapes. Figures joined with shapes at their edges or with a part(s) removed are composite figures. Below is a summary of the formulas you have worked with.

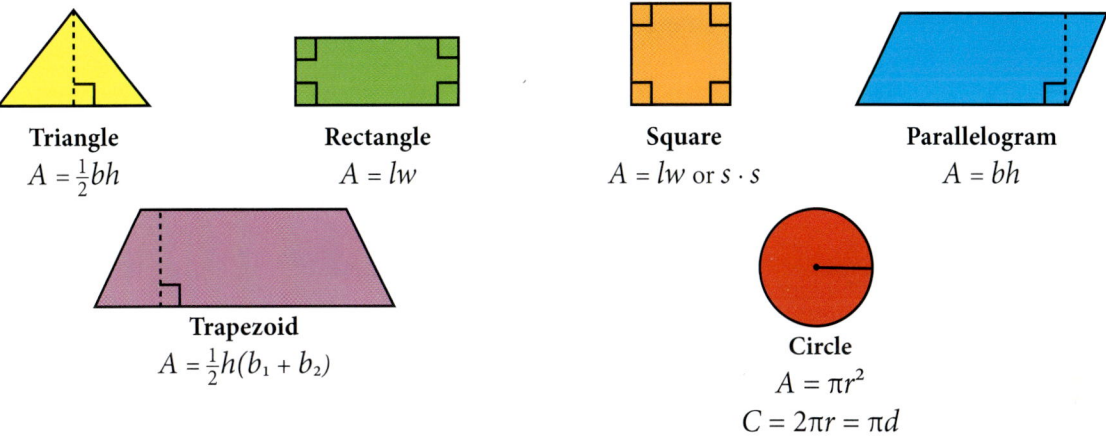

Triangle
$A = \frac{1}{2}bh$

Rectangle
$A = lw$

Square
$A = lw$ or $s \cdot s$

Parallelogram
$A = bh$

Trapezoid
$A = \frac{1}{2}h(b_1 + b_2)$

Circle
$A = \pi r^2$
$C = 2\pi r = \pi d$

The area of each shape in a composite figure must be determined in order to find the total area of the figure. Once the areas are calculated, decide whether to add or subtract the areas to find the overall area of the composite figure.

You can find the total area of the figure shown below by breaking it up into familiar shapes and adding the areas of the shapes together.

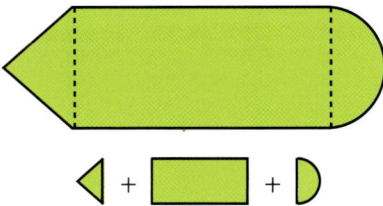

You can find the shaded area of the figure below by calculating the area of each shape and subtracting the smaller area from the larger area.

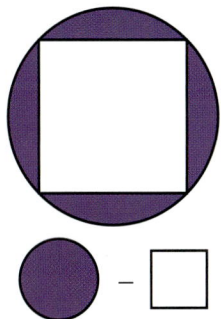

Lesson 2.7 ~ Composite Figures **65**

EXAMPLE 1

Calculate the area of the shaded region.

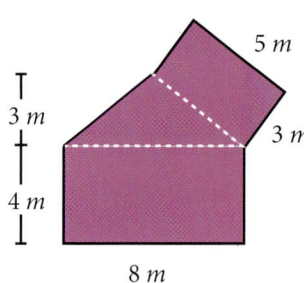

SOLUTION

Draw a diagram.

Find the area of each shape.

   Area = $lw = 8 \cdot 4 = 32\ m^2$

   Area = $\frac{1}{2}bh = \frac{1}{2}(8)(3) = 12\ m^2$

Area = $lw = 5 \cdot 3 = 15\ m^2$

Add the areas of the three shapes. $32 + 12 + 15 = 59\ m^2$

The area of the shaded region is 59 square meters.

EXAMPLE 2

Calculate the area of the shaded region. Use 3.14 for π.

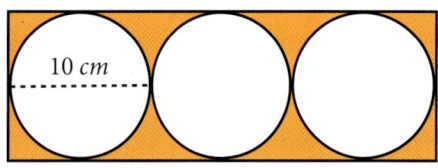

SOLUTION

Draw a diagram.

 −  −  −

Find the area of the rectangle.   Area = $lw = 10 \cdot 30 = 300\ cm^2$

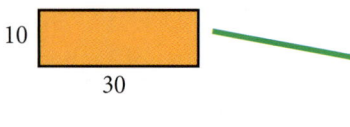

The length of the rectangle is three diameters and the width is one diameter.

Find the area of one circle.   Area = $\pi r^2 \approx (3.14)(5)^2 \approx 78.5\ cm^2$

Subtract the area of the three circles from
the area of the rectangle. $300 - 78.5 - 78.5 - 78.5 = 64.5\ cm^2$

The area of the shaded region is about $64.5\ cm^2$.

Lesson 2.7 ~ Composite Figures

EXERCISES

Use a diagram to show how to find the area of each shaded region.

1.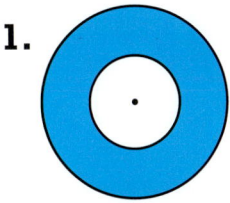
2.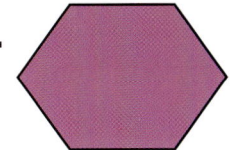
3.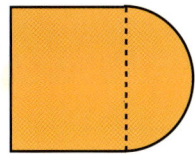
4.
5.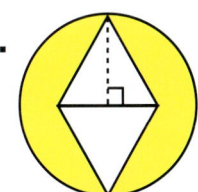
6.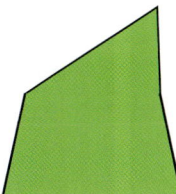

Calculate the area of each shaded region. Use 3.14 for π. If necessary, round to the nearest hundredth.

7.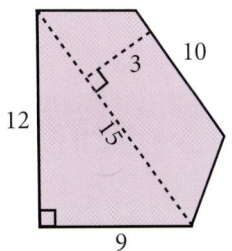
8.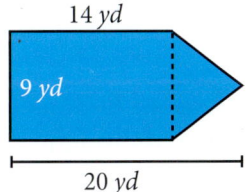
9. AB = 4 cm, AC = 6 cm
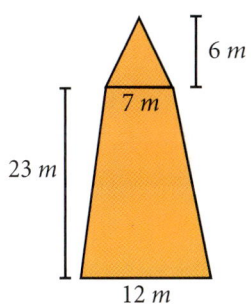

10.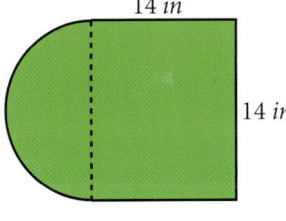
11.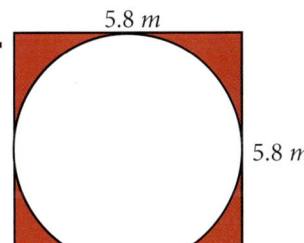
12.

13.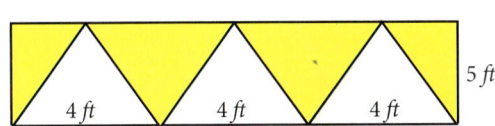
14.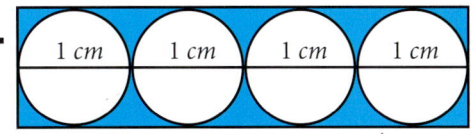

15. Keisha needed to find the area of the shaded region at right. Her work is shown below, but has an error. Identify the error and then find the area of the shaded region using 3.14 for π.

Keisha's Work
□ A = lw = (8)(8) = 64
○ A = πr² ≈ π(4)² ≈ 50.24
Total Area = 64 + 50.24 = 114.24 cm²

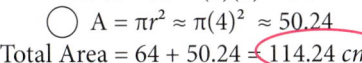

Lesson 2.7 ~ Composite Figures **67**

Find the perimeter of each composite figure. Use 3.14 for π.

16.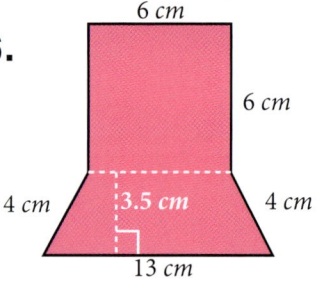

17.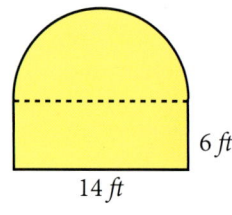

18.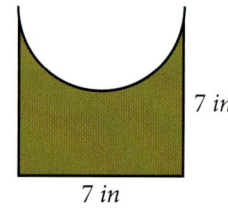

19. Tim drilled a 2 inch diameter hole in a sheet of metal. The metal sheet had an original area of 25 square inches. How much area remained? Use 3.14 for π.

20. The bases of a trapezoid are 13 *cm* and 8 *cm*. The height is 6 *cm*. Amir found the largest triangle that will fit inside the trapezoid.
   a. What were the dimensions for the triangle?
   b. He cut the triangle out of the trapezoid. How much area remained after the triangle was removed?

21. Sydney needs to cut rectangles from a 4 by 8 foot sheet of plywood. Each rectangle she cuts needs to be 1 foot by 3 feet. What is the maximum number of rectangles she can cut? Show all work necessary to justify your answer.

22. A rectangular track surrounds the edge of the football field at McKinley Middle School. The football field is 360 feet long and 160 feet wide. The inside edge of the 10 foot wide track borders the football field.
   a. Draw a diagram of the track and the football field.
   b. Find the total area of the track.

23. Claire sent a postcard to her grandmother. The card was 3 inches by 5 inches. She put a circular sticker on one side. The sticker had a radius of 1.25 inches. How much space is left on the side with the sticker for her to write on? Use 3.14 for π. Round to the nearest hundredth.

REVIEW

Solve for *x*.

24.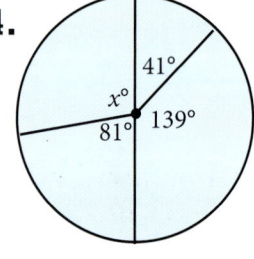

25.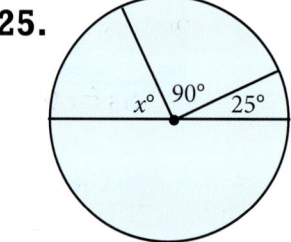

26.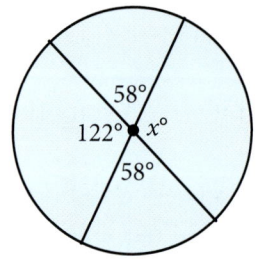

Plot each set of points. Connect in the order given. Find the area of each figure.

27. (3, 5), (3, −2), (−4, −2)

28. (0, −3), (8, −3), (8, 1), (0, 1)

CIRCLE SIMILARITY

LESSON 2.8

 Apply properties of similarity to circles.

Two figures with the exact same shape, but not necessarily the same size, are called similar figures. All circles have the exact same shape; therefore, all circles are similar. Similar figures can be compared using a ratio. A **ratio** is a comparison of two numbers using division. The ratio of *a* to *b* can be written three different ways.

$$\frac{a}{b} \quad a:b \quad a \text{ to } b$$

EXPLORE!   STEPPING STONES

Latrelle is putting stepping stones around his backyard. He bought three different sizes of circular stepping stones. He wants to compare the three sizes with ratios.

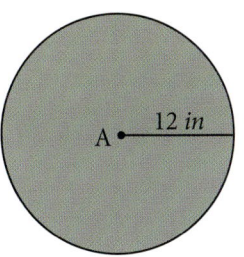 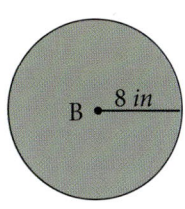 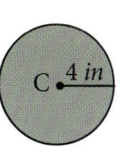 

Step 1: Write a ratio comparing the radius of one circle to the radius of another circle.

Write each ratio as a fraction in simplest form.
  a. ⊙A to ⊙C   **b.** ⊙B to ⊙C   **c.** ⊙A to ⊙B

Step 2: Find the **exact** circumference of each circle.

Step 3: Write a ratio comparing the **exact** circumference of one circle to the circumference of another circle. Write each ratio as a fraction in simplest form.
  a. ⊙A to ⊙C   **b.** ⊙B to ⊙C   **c.** ⊙A to ⊙B

Step 4: What do you notice about the ratios in **Step 1** compared to the ratios in **Step 3**?

Step 5: What do you predict about the ratios of the diameters for the three stepping stones?

Step 6: Latrelle buys one extra-large stepping stone with a circumference of 48π inches. How many times larger is this stepping stone's radius than that of ⊙C? Use words and/or numbers to show how you determined your answer.

Step 7: Find the **exact** area of each of Latrelle's stepping stones.

EXPLORE! (CONTINUED)

Step 8: Write a ratio comparing the area of one circle to the area of another circle. Write each ratio as a fraction in simplest form.

    a. ⊙A to ⊙C   **b.** ⊙B to ⊙C   **c.** ⊙A to ⊙B

Step 9: What do you notice about the ratios in **Step 8** compared to the original ratios in **Step 1**?

Step 10: Suppose Latrelle's one extra-large stepping stone has a radius 6 times as large as the radius of his smallest stone. How many times larger is the area of the extra-large stone compared to the smallest stone?

CIRCLE SIMILARITY

If the ratio of the radii of two circles is $a : b$, then:
- The ratio of their diameters is $a : b$.
- The ratio of their circumferences is $a : b$.
- The ratio of their areas is $a^2 : b^2$.

EXAMPLE 1   Two circles have diameters of 12 *cm* and 4 *cm*. What is the ratio of their circumferences?

SOLUTION

Write a ratio comparing the diameters. $\dfrac{12\ cm}{4\ cm}$

Simplify. $\dfrac{12\ cm}{4\ cm} = \dfrac{3}{1}$

The circumferences have the same ratio as the diameters. The ratio of their circumferences is 3 : 1.

EXAMPLE 2   Two circles have circumferences of 24π and 36π. Find the ratio of their areas.

SOLUTION

Write a ratio comparing their circumferences. $\dfrac{24\pi}{36\pi}$

Simplify. $\dfrac{24\pi}{36\pi} = \dfrac{24}{36} = \dfrac{2}{3}$

Square the ratio to find the ratio of their areas. $\dfrac{2}{3} \rightarrow \dfrac{2^2}{3^2} = \dfrac{4}{9}$

The ratio of their areas is 4 : 9.

A **proportion** is an equation stating two ratios are equivalent. Proportions can be used to find missing measures between two circles. Proportions can be solved using cross products.

Proportion	Cross Products	The cross products of a proportion are equal.
$\dfrac{3}{5} = \dfrac{12}{20}$	$3(20) = 5(12)$	$60 = 60$

EXAMPLE 3   Use a proportion to find the length of the radius in ⊙Y.

C = 78.5 in   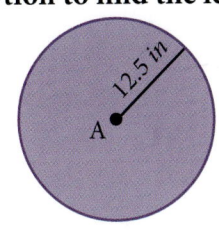   C = 15.7 in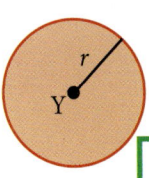

SOLUTION

Write a ratio comparing the circumferences. $\frac{78.5}{15.7}$

Write a ratio comparing the radii. $\frac{12.5}{r}$

Set the two ratios equal to each other. $\frac{78.5}{15.7} = \frac{12.5}{r}$

Set the cross products equal to each other. $78.5r = (15.7)(12.5)$

Multiply, then divide on both sides. $\frac{78.5r}{78.5} = \frac{196.25}{78.5}$

   $r = 2.5$

The radius of ⊙Y is 2.5 in.

If the larger circle's circumference is in the numerator of the first ratio, make sure the radius of the larger circle is in the numerator of the second ratio.

EXERCISES

Write a ratio comparing the radii of each pair of circles. Write the ratio in simplest form.

1.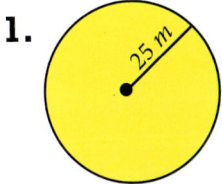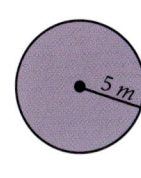

2. C = 24π, C = 6π

3.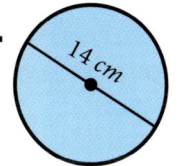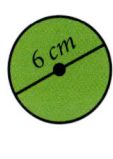

4. ⊙P has a diameter of 52 yards. ⊙W has a diameter of 52 yards. Write a ratio comparing the circles' radii.

5. The radius of ⊙A is 4 in. The radius of ⊙B is 22 in. Write a ratio comparing the circles' circumferences.

6. The circumference of one circle is 100 miles. The circumference of another circle is 75 miles. Write a ratio comparing the circles' diameters.

7. A car tire has a diameter of 30.4 inches. A tractor tire has a diameter of 76 inches. Write a ratio comparing the circumferences of the tires. Write the ratio as a fraction without decimals and in simplest form.

Write a ratio comparing the areas of each pair of circles. Write the ratio in simplest form.

8.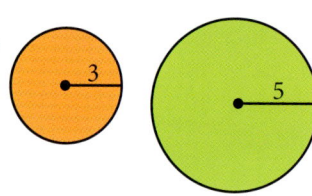

9. C = 100 cm, C = 400 cm

10.

11. ⊙A has a diameter of 32 feet. ⊙M has a diameter of 24 feet. Write a ratio comparing the circles' areas.

Use a proportion to find each missing measure.

12. C = 50.24 units   C = 6.28 units

13. C = 56.52 in   C = 37.68 in
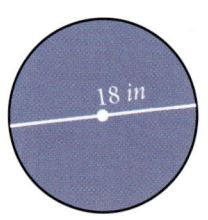 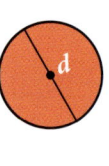

14. The circumference of ⊙A is 100.48 inches. The diameter is 32 inches. The circumference of ⊙B is 113.04 inches. Find the diameter of ⊙B.

15. The radii of two circles have a ratio of 2 : 5. The area of the larger circle is 125 square meters. Find the area of the smaller circle.

16. The diameters of two circles have a ratio of 3 : 4. The area of the smaller circle is 162 square units. Find the area of the larger circle.

17. The circumference of one circle is five times as large as the circumference of another circle. The radius of the smaller circle is 7.1 cm. Find the length of the larger circle's radius.

18. Jackson is shopping for a circular swimming pool. The Water Mania pool has a radius of 10 feet. The circumference is 62.8 feet. The circumference of the Splash Attack pool is 81.64 feet.
   a. Find the radius of the Splash Attack pool.
   b. Jackson's backyard has room for a pool with a radius of 12 feet. Jackson tells his mother the Splash Attack pool will fit perfectly. Do you agree or disagree? Explain your reasoning.

19. Explain why all circles are similar.

Saheel's Work

Ratio of diameters: $\frac{12}{28} = \frac{3}{7}$

Ratio of areas: $\frac{3^2}{7^2} = \frac{9}{49}$

Area proportion: $\frac{9}{49} = \frac{A}{615.44}$

49A = 5538.96
A = 113.04 cm²

20. The diameters of two circles are 12 cm and 28 cm. The area of the large circle is about 615.44 cm².
   a. Find the ratio of the radii.
   b. Saheel used a proportion to find the area of the small circle. His work is correctly shown to the left. Show another way to find the area of the small circle.

REVIEW

Find the area of each shaded region. Use 3.14 for π. Round to the nearest hundredth, if necessary.

21.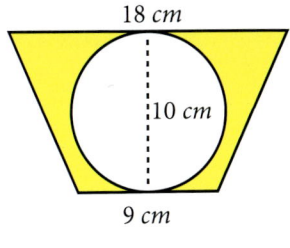

22.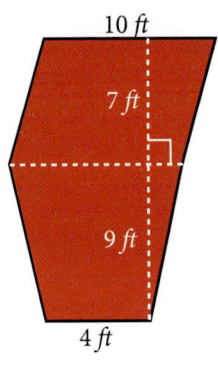

23.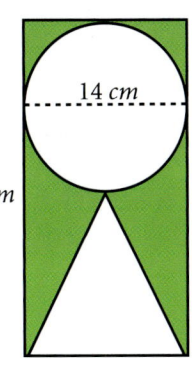

24.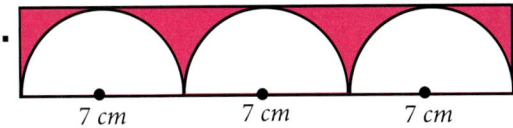

25.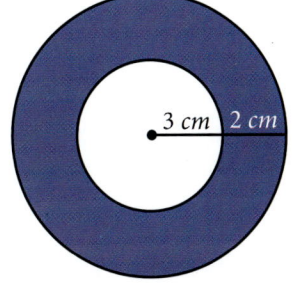

Tic-Tac-Toe ~ Circles and Squares

1. Find the area of the shaded region in each figure. The length of each side of the square is 10 cm. Use 3.14 for π.

   a. 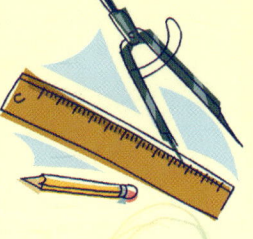   b.    c.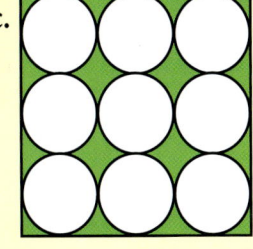

2. How many circles will be in each of the next two figures if the pattern is continued?

3. Sketch diagrams of the next two figures. Shade to match **figures a, b and c**.

4. Calculate the area of the "shaded" region in your new diagrams.

5. What percentage of the area is shaded in each figure? Do you see a pattern? Explain your reasoning.

Lesson 2.8 ~ Circle Similarity

AREA OF SECTORS

LESSON 2.9

 Use proportions to calculate the area of sectors.

A **sector** is a portion of a circle contained by two radii. A sector is sometimes referred to as a slice of a circle. A sector is described by the measure of its central angle. Remember that the sum of all central angles in a circle is 360°.

⊙C has a sector that is 90°. Ninety degrees is a fraction of the whole circle, which is 360°. This ratio can be used in a proportion to find the area of the sector. The ratio of the central angle to 360° can be set equal to the ratio of the area of the sector to the area of the circle.

$$\frac{90°}{360°} = \frac{\text{area of sector}}{\text{area of } \odot C}$$

AREA OF A SECTOR

$$\frac{\text{degree of the sector}}{360°} = \frac{\text{area of the sector}}{\text{area of the circle}}$$

EXAMPLE 1 The area of ⊙A is 60 cm^2. Use a proportion to find the area of the sector.

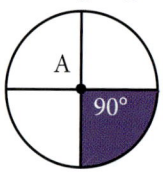

SOLUTION

Write a proportion.	$\frac{90°}{360°} = \frac{x}{60}$
Set the cross products equal to each other.	$90(60) = 360x$
Multiply.	$5400 = 360x$
Divide both sides of the equation by 360.	$\frac{5400}{360} = \frac{360x}{360}$
	$15 = x$

The area of the sector is 15 cm^2.

EXAMPLE 2   The Macy family bought a family-sized chocolate chip cookie for dessert. Erin ate a slice of the cookie with a central angle of 37°. The entire cookie had an area of 249 square centimeters. Find the area of Erin's slice to the nearest hundredth.

SOLUTION

Write a proportion.
$$\frac{37°}{360°} = \frac{x}{249}$$

Set the cross products equal to each other.
$$37(249) = 360x$$

Multiply.
$$9213 = 360x$$

Divide both sides of the equation by 360.
$$\frac{9213}{360} = \frac{360x}{360}$$

$$25.59 \approx x$$

The area of Erin's slice of cookie was approximately 25.59 square centimeters.

EXAMPLE 3   Find the area of the shaded sector in ⊙M. Round to the nearest hundredth.

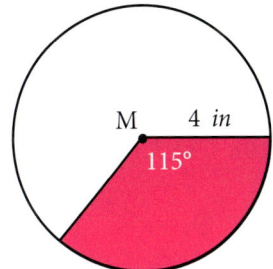

SOLUTION

Find the area of the circle.
$$\text{Area} = \pi r^2$$
$$\text{Area} \approx (3.14)(4)^2$$
$$\text{Area} \approx 50.24 \text{ square inches}$$

Substitute all known values into a proportion.
$$\frac{115°}{360°} \approx \frac{x}{50.24}$$

Set the cross products equal to each other.
$$(115)(50.24) \approx 360x$$

Multiply.
$$5777.6 \approx 360x$$

Divide both sides of the equation by 360.
$$\frac{5777.6}{360} \approx \frac{360x}{360}$$

$$16.04\overline{8} \approx x$$

Round to the nearest hundredth.
$$16.05 \approx x$$

The area of the shaded sector is approximately 16.05 square inches.

EXERCISES

Find the area of each shaded sector. Use 3.14 for π. Round to the nearest hundredth, when necessary.

1. A = 50 cm²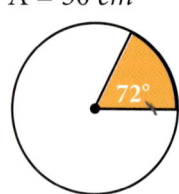

2. A = 28 in²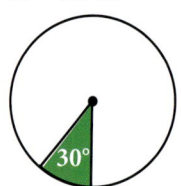

3. A = 46.5 ft²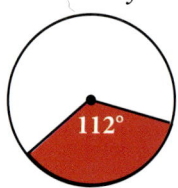

4. The area of a circle is 50.24 square miles. What is the area of a sector with a 60° central angle?

5. The radius of a circle is 3 inches. Find the area of a sector with a 35° central angle. Use 3.14 for π.

6. A sprinkler rotates back and forth watering a sector with a 75° central angle. The sprinkler sprays water a distance of 50 feet. Find the area the sprinkler waters. Use 3.14 for π.

7. The hour hand on a living room clock is 4 inches long. Use 3.14 for π.
   a. What is the central angle of the hour hand from noon to 9 p.m.?
   b. What is the area of the clock that is crossed over by the hour hand from noon to 9 p.m.? Use mathematics to justify your answer.

Find the area of each shaded sector. Use 3.14 for π. Round to the nearest hundredth, when necessary.

8.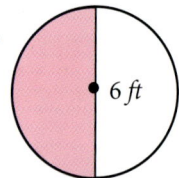

9.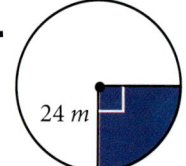

10.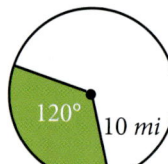

11.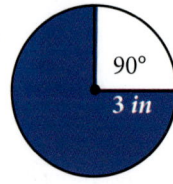

12.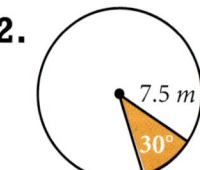

13.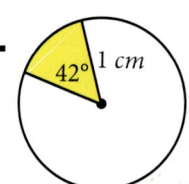

14. Determine in which situation Isaiah will get the most to eat. Show all work necessary to justify your answer.
    Situation 1: A pizza cut into 8 pieces has a diameter of 14 inches. Isaiah eats 5 pieces.
    Situation 2: A pizza with a diameter of 18 inches is cut into 12 pieces. Isaiah eats 4 pieces.

15. The area of a circle is 36π yd². Find the exact area of a sector with a central angle of 30°.

16. Chris says every circle with a shaded sector of 180° will have the same area for the shaded sector. Do you agree or disagree? Explain your reasoning.

17. A family ate all but one slice of pie. The remaining slice of pie has a central angle of 60°. The pie pan has a diameter of 8 inches. Find the area of the pie the family ate.

18. Find the area of sector 1 in ⊙B.

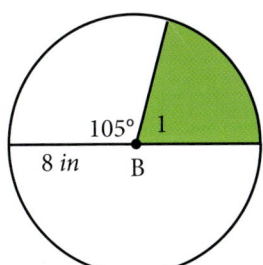

19. Find the area of sector 1 in ⊙T.

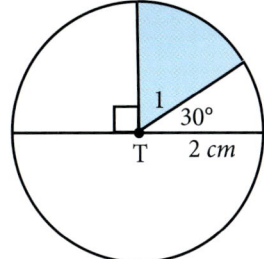

REVIEW

20. LaTisha's swimming pool has a circumference of 31.4 feet. What is the length of the radius of the pool?

21. The circumferences of two circles have a ratio of 2 : 3. The diameter of the larger circle is 7.5 units. Find the diameter of the smaller circle.

22. The radii of two circles have a ratio of 2 : 5. The area of the smaller circle is 350 meters.
 a. Find the ratio of the areas of the circles.
 b. Find the area of the larger circle.

Tic-Tac-Toe ~ Pie Charts

A pie chart is a data display using sectors of a circle. The pie chart below shows the results of a survey of high school students about their favorite subject in school.

1. Use a proportion or the percent equation to determine the measure of each central angle in the pie chart. Remember that the sum of all central angles in a circle is 360°.

2. Measure the length of the radius of the pie chart using centimeters. Find the area of each sector. Use 3.14 for π. Round to the nearest hundredth. Label each answer based on the subject that sector represents on the chart.

3. Another pie chart has an area of 314 square centimeters. Find the central angles of each of the four regions given the area of their sectors.

Favorite Sport to Watch
Football   131.88 cm^2
Basketball   87.92 cm^2
Baseball   53.38 cm^2
Soccer   40.82 cm^2

REVIEW

BLOCK 2

Vocabulary

area	circumference	proportion
center	composite figures	radius
central angle	diameter	ratio
chord	perimeter	sector
circle	pi	trapezoid

Use area formulas to compute area of geometric shapes and find missing measures.
Understand and use the trapezoid area formula.
Identify, name and define parts of circles.
Understand and use the relationship between pi and diameter to find circumference.
Understand and use the circle area formula.
Identify and use common estimations of pi.
Find the area of composite figures.
Apply properties of similarity to circles.
Use proportions to calculate the area of sectors.

Lesson 2.1 ~ Areas of Triangles and Parallelograms

Find the area of each figure.

1.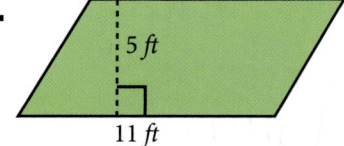
2.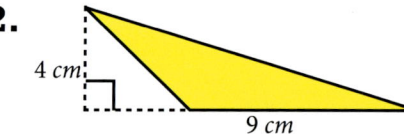
3.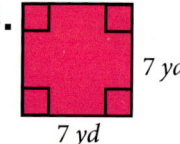

4. Jacklyn has a triangle pendant with a base of 9 *mm* and a height of 5 *mm*. What is the area of the pendant?

5. A square has a perimeter of 24 feet. What is the area of the square?

Find each missing measure. Show all work necessary to justify your answer.

6. $A = 30\ ft^2$

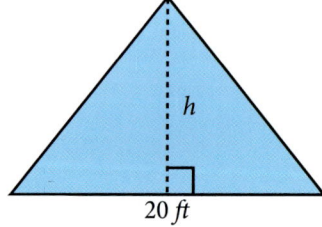

7. $A = 18\ m^2$

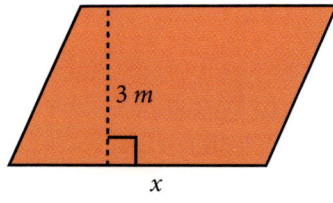

8. $A = 23.4\ in^2$

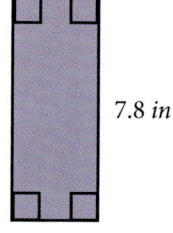

Lesson 2.2 ~ Area of a Trapezoid

Find the area of each trapezoid.

9.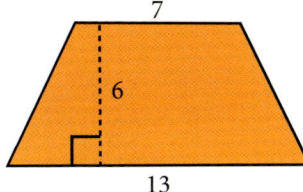

10.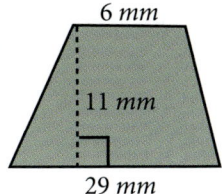

11. Base measures are 5.7 *cm* and 8.3 *cm*. The height of the trapezoid is 9 *cm*.

Find the unknown base or height of each trapezoid. Show all work necessary to justify your answer.

12. A = 45 square inches
b_1 = 7 inches
b_2 = 11 inches
h = ?

13. A = 160 in^2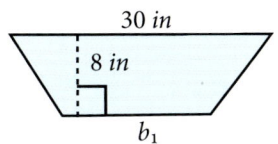

14. A = 76 m^2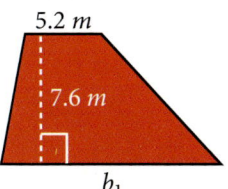

15. A brick of gold looks like a trapezoid from a side view. The length of the top is 3.5 inches. The bottom base measures 7 inches. The area of a side of the trapezoid is approximately 17.5 square inches. What is the height?

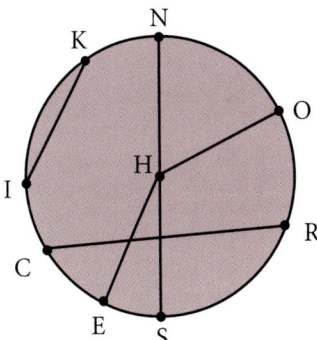

Lesson 2.3 ~ Parts of a Circle

Use ⊙H to name each of the following.

16. the center

17. two radii

18. a diameter

19. two chords

20. two central angles

21. the longest chord

Solve for *x*.

22.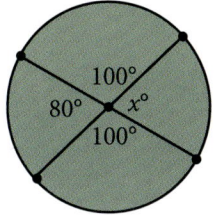

23.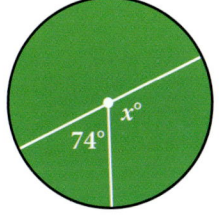

24.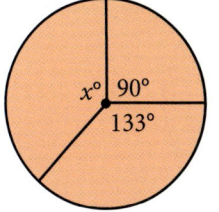

25. Draw ⊙P with a diameter $\overline{TV}$ and chord $\overline{TM}$.

Lesson 2.4 ~ Circumference and Pi

Find the circumference of each circle. Use 3.14 for π.

26.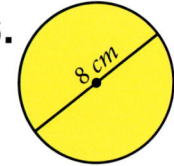

27.

28. The radius of a bicycle tire is 25.4 *cm*. What is the circumference of the tire?

29. The first Ferris wheel had a diameter of 250 feet. How far did a person travel during one revolution on the Ferris wheel?

Find each missing measure. Use 3.14 for π.

30. Rudy ran laps on a circular track. He ran 314 meters in one lap. Find the diameter of the track.

31. Samantha made a circular cake for her brother. The circumference of the cake is 50.24 inches. Find the approximate radius of the cake.

32. C = 75.36 *cm*
    d ≈ ?
    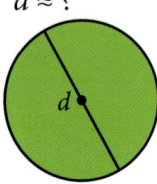

33. C = 25.12 *yd*
    r ≈ ?

34. C = 329.7 *in*
    r ≈ ?
    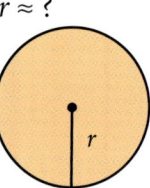

Lesson 2.5 ~ Area of a Circle

Find the area of each circle. Use 3.14 for π.

35. a circle with a radius of 3 centimeters

36. a circle with a diameter of 24 *mm*

37.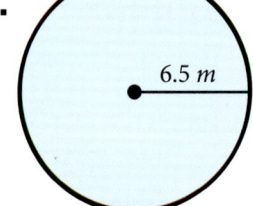

38.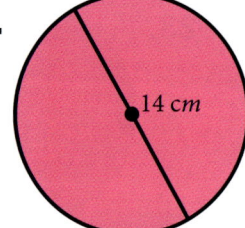

39. The diameter of a circle is 80 feet. Find the area of the circle.

40. A flying disk has a radius of 6 inches. Find the area of the disk.

Calculate the exact area of each circle.

41.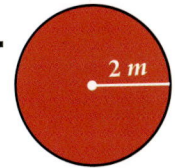

42.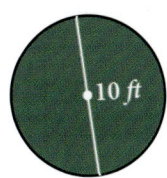

43. Pierre found the exact area of a circle with a diameter of 8 *ft*. His work is at right. Identify Pierre's error and find the exact area of the circle.

Pierre's Work
$A = \pi r^2$
$A = \pi(8)^2$
$A = 64\pi \; ft$

44. How many times larger is a circle with a radius of 2 *in* than a square with a side length of 2 *in*? Use words and/or numbers to show how you determined your answer.

Lesson 2.6 ~ More Pi

Identify the most appropriate form of π to use in each situation. Explain your choice. Find each missing measure.

45.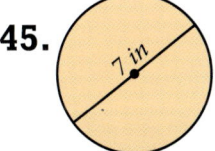

Circumference = ?

46.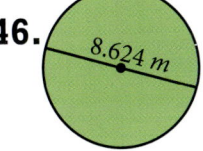

Area = ?

47.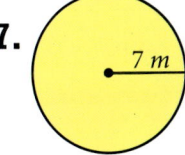

Area = ?

48. A fish bowl has a circular base with a radius of 14 inches. Calculate a rough estimate for the area of the base of the fish bowl.

49. Terry wants the most accurate answer possible when finding the area of her circular quilt. It has a radius of 4.7 feet.
   a. Should she use 3.14, $\frac{22}{7}$ or the π button on her calculator? Explain your reasoning.
   b. Find the area of the quilt using your answer from **part a**. Round the answer to the nearest hundredth.

50. Give an example of a real-world situation where it would be appropriate to use 3 as an estimate of π.

51. Give an example of a real-world situation when an exact answer may be needed to find the area or circumference of a circle.

Lesson 2.7 ~ Composite Figures

Calculate the area of each shaded region.

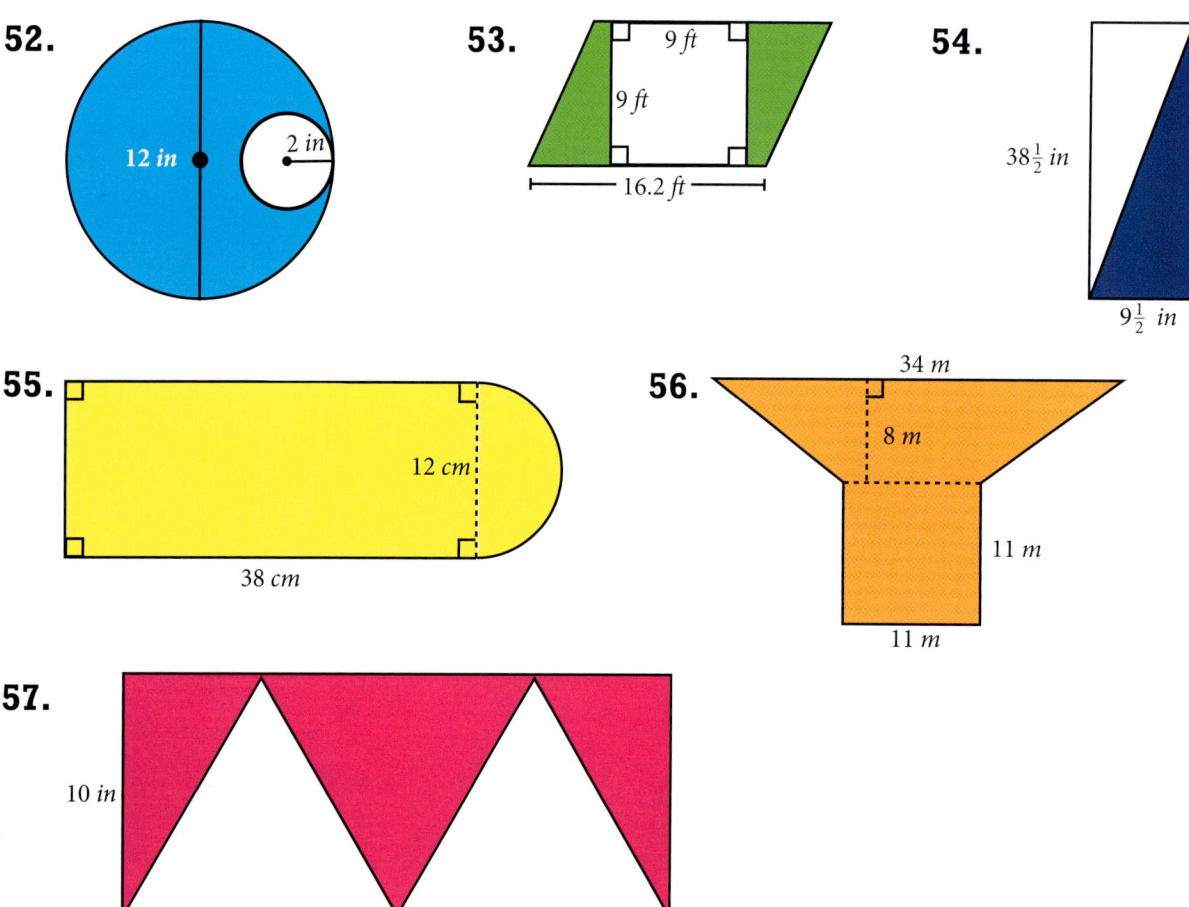

52. Circle with diameter 12 in, small circle with radius 2 in removed.

53. Parallelogram with base 16.2 ft containing a 9 ft × 9 ft square; green regions on sides.

54. Rectangle $38\frac{1}{2}$ in by $9\frac{1}{2}$ in with shaded triangle.

55. Yellow figure: rectangle 38 cm × 12 cm with semicircle on end.

56. Orange figure with top width 34 m, height 8 m, lower rectangle 11 m × 11 m.

57. Rectangle 10 in by (12 in + 12 in) with two white triangles removed.

Lesson 2.8 ~ Circle Similarity

Write a ratio comparing the radii of each pair of circles. Write the ratio in simplest terms.

58. Circles with radii 10 and 2.

59. Circles with diameter 24 m and diameter 60 m.

60. Grace has two flower pots. The base of one has a circumference of 16 inches. The base of the other flower pot has a circumference of 10 inches. Write a ratio comparing the pots' diameters.

Write a ratio comparing the areas of each pair of circles. Write the ratio in simplest form.

61. Circles with diameter 10 ft and diameter 5 ft.

62. Circles with radii 2 and 5.

63. ⊙B has a circumference of 72 meters. ⊙X has a circumference of 18 meters. Write a ratio comparing the areas of the circles.

Use a proportion to find each missing measure. Round to the nearest hundredth, as needed.

64. C = 21.98 in   C = ?

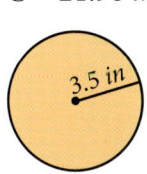 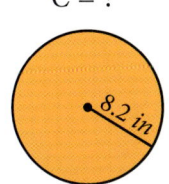

65. C = 46 cm   C = 34.5 cm

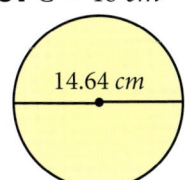 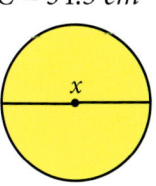

66. The radii of two circles have a ratio of 4 : 5. The larger circle has an area of 1000 square centimeters.
    a. What is the ratio of the areas of the circles?
    b. Use the ratio in **Part a** and the area of the largest circle to write a proportion.
    c. Find the area of the smaller circle.

67. The radii of two circular gears have a ratio of 2 : 7. The area of the small gear is 8 square inches. Find the area of the large gear.

Lesson 2.9 ~ Area of Sectors

Use a proportion to find the area of each shaded sector. Round answers to the nearest hundredth, if necessary.

68. A = 58 cm²

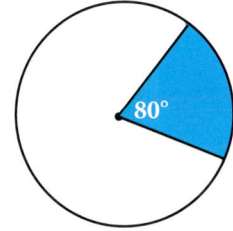

69. A = 254.34 in²

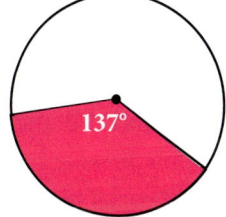

70. A circle has an area of 278.9 square meters. Find the area of a sector with a central angle of 66°.

71. A pumpkin pie has an area of 113.04 square inches. The pie is cut into 8 equal slices. Find the area of one slice.

72. The exact area of a circle is 78π square miles. Find the exact area of a sector with a central angle of 120°.

73. A circular waffle has an area of 200.96 square centimeters. Find the area of a slice that has a central angle of 90°.

74. The radius of a circle is 5 inches. Find the area of a sector of the circle with a central angle of 200°. Use 3.14 for π.

CAREER FOCUS

Manish
Safety and Health Professional

I am an occupational safety and health professional. I ensure that my company is meeting safety and health codes, conducting safety trainings and completing inspections. If an accident happens, I investigate it and handle any workers' compensation claims. Most importantly, I make sure the workplace is a safe and healthy place for everybody who works there.

I use math to help communicate injury prevention to managers and employees of all levels. For example, I calculate incident rates (the number of injuries or illnesses per 100 employees). I also determine injury causes and calculate percentages by type of injury and body part. Sometimes I analyze chemical and noise exposure levels to make sure that people are not exposed to harmful contaminants. Math helps me know that the workplace is as safe as possible.

Many safety professionals are required to get a specialized degree or certification. I have a Bachelor's of Science in Occupational Safety and Health degree from Oregon State University. Some companies require special certifications that can be obtained by attending extra training classes or passing certain exams.

The median salary of an occupational health and safety specialist is approximately $54,920 per year. The middle 50 percent earned between $41,800 and $70,230 per year. The lowest 10 percent earned less than $32,230 per year and the highest 10 percent earned more than $83,720 per year. Health and safety specialists can work for private industries, hospitals or governments. Salaries vary depending on which type of employer one works for.

I like being a safety and health professional because I can use math, psychology and business. I am constantly balancing the three subjects and really enjoy the diversity. I'm also proud to know that I am doing my part in keeping employees safe and healthy at work.

CORE FOCUS ON SHAPES & ANGLES
BLOCK 3 ~ SURFACE AREA AND VOLUME

Lesson 3.1	Three-Dimensional Figures	87
Lesson 3.2	Drawing Solids	92
	Explore! Netting A Solid	
Lesson 3.3	Slicing Solids	96
	Explore! Cutting Clay	
Lesson 3.4	Surface Area of Prisms	101
	Explore! Take Your Pick	
Lesson 3.5	Volume of Prisms	106
	Explore! Cutting Corners	
Lesson 3.6	Surface Area of Regular Pyramids	111
	Explore! Tent Making	
Lesson 3.7	Volume of Pyramids	115
	Explore! Pyramid vs. Prism	
Review	Block 3 ~ Surface Area and Volume	119

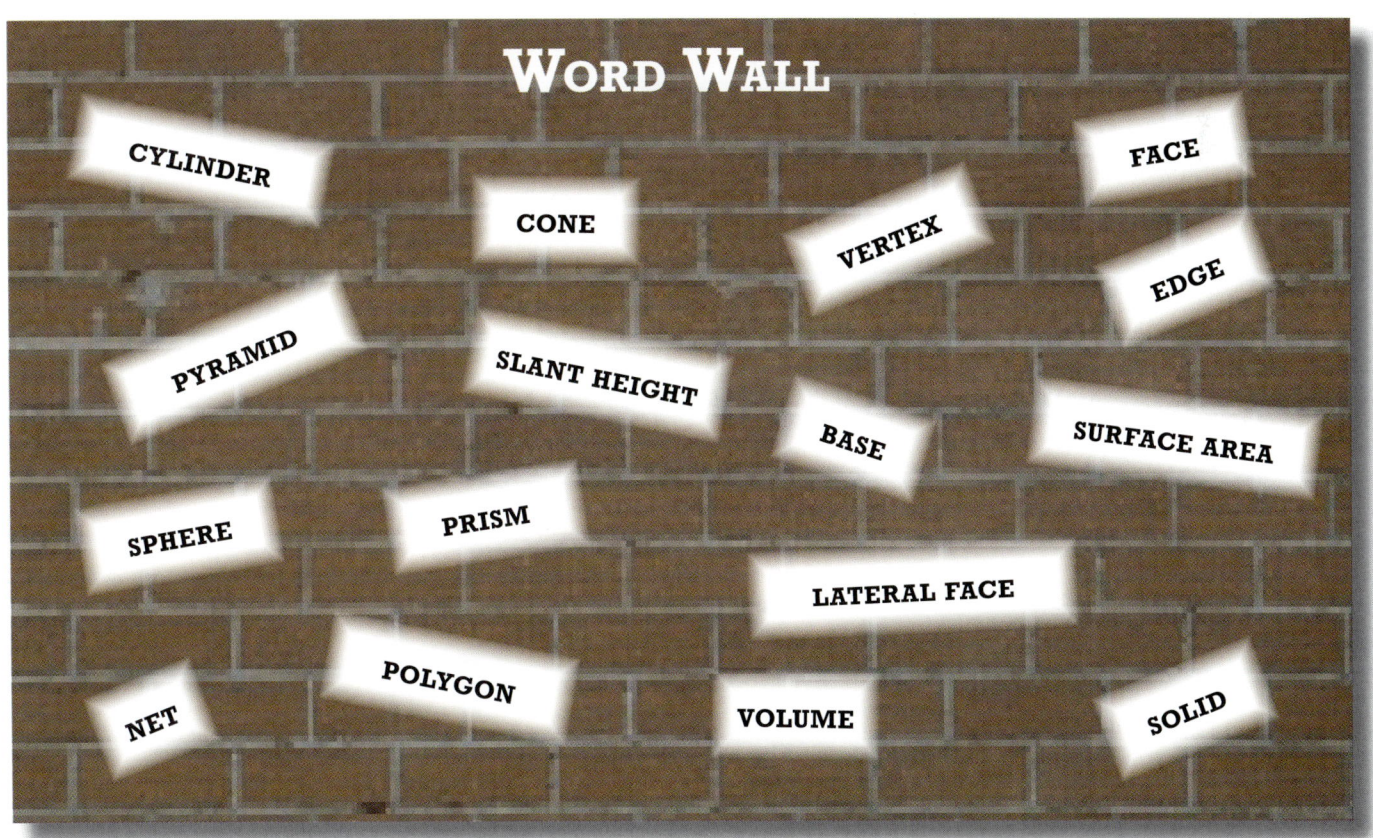

BLOCK 3 ~ SURFACE AREA AND VOLUME
TIC-TAC-TOE

STENCILS

Create a set of stencils that can be used to make a set of solids.

See page 105 for details.

NETTING A CUBE

Draw multiple nets for the same prism.

See page 95 for details.

A SOLID VIEW

Draw different views of three dimensional figures.

See page 100 for details.

NOT REGULAR

What does it mean to be a regular geometric solid? Find the surface area of solids that are not regular.

See page 123 for details.

EULER'S FORMULA

Discover Euler's Formula relating the number of faces, vertices and edges in a solid.

See page 91 for details.

A SOLID ALBUM

Make a photo album of solids.

See page 118 for details.

RESEARCH SAYS...

Research a landmark or building. Calculate the surface area.

See page 95 for details.

MORE CEREAL PLEASE

Compare the volume of cereal boxes to the weight of the cereal.

See page 110 for details.

VOLUME OF A SPHERE

Find the volume of various spherical objects.

See page 105 for details.

THREE-DIMENSIONAL FIGURES

LESSON 3.1

 Identify the names and qualities of three-dimensional figures.

The world is made up of many three-dimensional objects. From the blocks children play with to the Great Pyramid of Egypt, solids are all around you. A **solid** is a three-dimensional figure that encloses a part of space. In this lesson you will learn about five different types of solids. Many solids are made up of flat surfaces called **faces**. Each face is a **polygon**, or a closed figure made up of three or more line segments. Solids may also have one or two **bases**, often located on the top or bottom of a solid. The bases are either polygons or circles.

Name	Definition	Diagrams
Prism	A solid formed by two congruent, parallel bases and rectangular sides.	
Pyramid	A solid with a polygonal base and triangular sides that meet at a vertex.	
Cylinder	A solid formed by two congruent and parallel circular bases.	
Cone	A solid formed by one circular base and a curved surface which connects the base and the vertex.	
Sphere	A solid formed by a set of points in space that are the same distance from a center point.	

Lesson 3.1 ~ Three-Dimensional Figures

Prisms and pyramids are named by the shape of their base(s). Each prism or pyramid has two parts to its name. The first part of the solid's name describes the shape of the base. The second part of its name describes the shape as either a prism or a pyramid.

> **NAMING A PRISM OR PYRAMID**
> 1. Describe the shape of the solid's base(s).
>   ◆ 3 sides = Triangular
>   ◆ 4 sides = Square, Rectangular or Trapezoidal
>   ◆ 5 sides = Pentagonal
>   ◆ 6 sides = Hexagonal
>   ◆ 7 sides = Heptagonal
>   ◆ 8 sides = Octagonal
> 2. Describe the shape as either a prism or pyramid.

Pentagonal prism 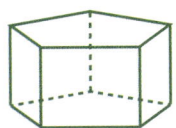 Triangular prism  Hexagonal pyramid 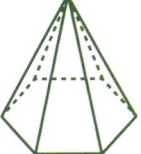 Square pyramid  Rectangular prism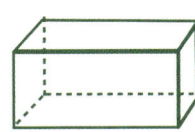

EXAMPLE 1 Name the solid that best describes each picture.

a.  b. 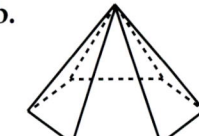 c.  d.

SOLUTIONS

a. The bases are the polygons parallel to one another. The bases are triangles. The solid is called a <u>triangular prism</u>.
b. The solid has one base with six sides. It is a <u>hexagonal pyramid</u>.
c. The solid has parallel circular bases. It is a <u>cylinder</u>.
d. The prism has parallel bases that are squares or rectangles. It can be called a <u>square prism</u> or <u>rectangular prism</u>.

There are two types of faces on a solid: bases and lateral faces. A base is usually the top or bottom of a solid. A **lateral face** is any side of a solid which is a polygon. An **edge** of a solid is a line segment where two faces meet. A **vertex** is a point where three or more edges meet. Examples of faces, vertices and edges are shown below.

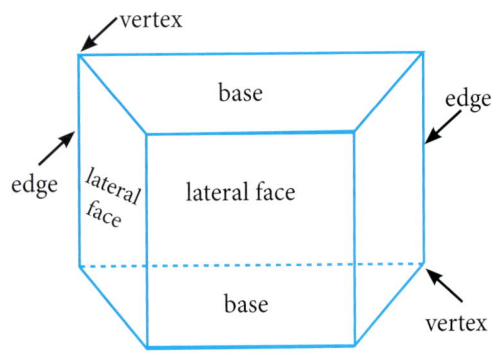
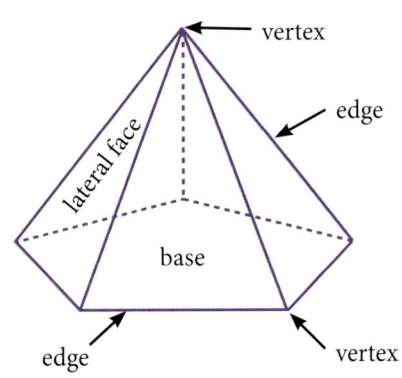

Lesson 3.1 ~ Three-Dimensional Figures

EXAMPLE 2 Identify the number of faces, vertices, edges, bases and lateral faces in the hexagonal pyramid.

SOLUTIONS
Bases: The polygon at the bottom of the prism. There is 1 base.
Lateral Faces: All polygonal sides that are not bases. There are 6 lateral faces.
Edges: The line segments formed when two faces meet. There are 12 edges.
Vertices: All corners where the edges meet. There are 7 vertices.
Faces: All polygonal sides including the base and all lateral faces. There are 7 faces.

EXERCISES

Name the solid that best describes each picture.

1.

2.

3.

4.

5.

6.

7.

8.

9.

Tell whether each statement is true or false. If false, explain your reasoning.

10. A cone has two bases.

11. A cylinder is more like a prism than a pyramid.

12. A prism has two bases that are not congruent.

13. The faces of a pyramid are rectangles.

14. The faces of a prism are rectangles.

15. A cone has one vertex.

16. A pyramid has one vertex.

17. All prisms have 12 vertices.

18. A prism has one base.

19. A pyramid is named by the shape of its lateral faces.

Identify the number of lateral faces, bases, edges and vertices.

20. 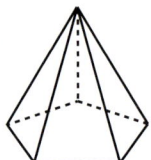   **21.** 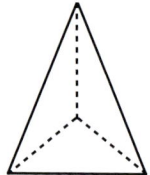   **22.**

23. Complete the table.

Name of solid	a.	e.	i.	m.
Number of faces (include base)	b.	f.	j.	n.
Number of vertices	c.	g.	k.	o.
Number of edges	d.	h.	l.	p.

24. Describe how to determine which faces are lateral faces and which faces are bases on a prism or pyramid.

25. How are the number of bases and lateral faces related to the total number of faces on a prism?

26. Explain why cylinders and cones do not have faces and edges.

Name a solid that fits each description.

27. a can of soda pop

28. a shoe box

29. a pyramid with four faces

30. a prism with eight lateral faces

31. a pyramid with six vertices

32. a basketball

33. Kiernan noticed that a triangular prism and a pentagonal pyramid both have 6 vertices and 6 faces. However, they do not have the same number of edges. Name two other pairs of prisms and pyramids that match using Kiernan's description of having the same number of vertices and the same number of faces. Explain how you know your answers are correct.

34. Azariah has two pentagonal pyramids equal in size. He wants to glue the two bases together to form one three-dimensional figure. He tells his friend that the new figure will have twice the number of vertices, faces and edges as one pentagonal pyramid. Is Azariah correct? Show all work necessary to justify your answer.

90   Lesson 3.1 ~ Three-Dimensional Figures

REVIEW

Find the circumference and area of each circle. Choose the most appropriate estimate of π to use.

35. 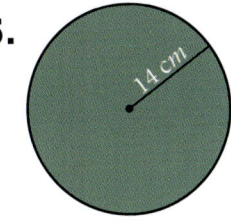 14 cm

36. 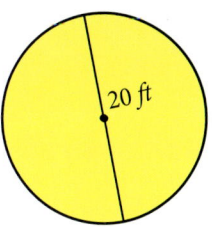 20 ft

37. 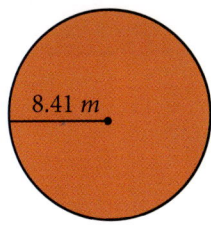 8.41 m

38. Clara and Beatrice made a circular baby blanket. It has a radius of 24 inches. Find the area of the baby blanket. Use 3.14 for π.

Tic-Tac-Toe ~ Euler's Formula

 Leonard Euler (1707-1783) proved many theorems and developed many formulas in his life. One formula he developed relates the number of edges, faces and vertices of a solid consisting of faces that are polygons (called polyhedra).

Step 1: Copy and complete the table below.

Solid	Vertices (V)	Faces (F)	Edges (E)
Cube			
Square Pyramid			
Pentagonal Prism			
Hexagonal Pyramid			

Step 2: Add a column to the table titled V + F − E. Find the value of V + F − E for each solid.

Step 3: Did you notice anything about your results in **Step 2**? Can you make any predictions about the formula Euler developed?

Step 4: Look up Euler's formula for polyhedra on the internet or another resource. Were you correct? State his formula.

Step 5: Will the formula hold true if you cut off the corner of a prism like the solid below? Show all work necessary to justify your answer.

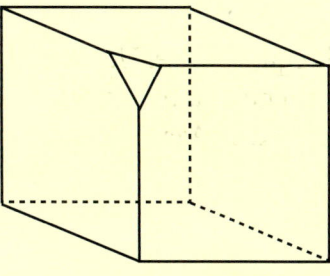

DRAWING SOLIDS

LESSON 3.2

 Sketch and draw solids in two dimensions and in three dimensions.

A **net** is a two-dimensional pattern that can be folded to form a three-dimensional figure. Below are some nets of solids.

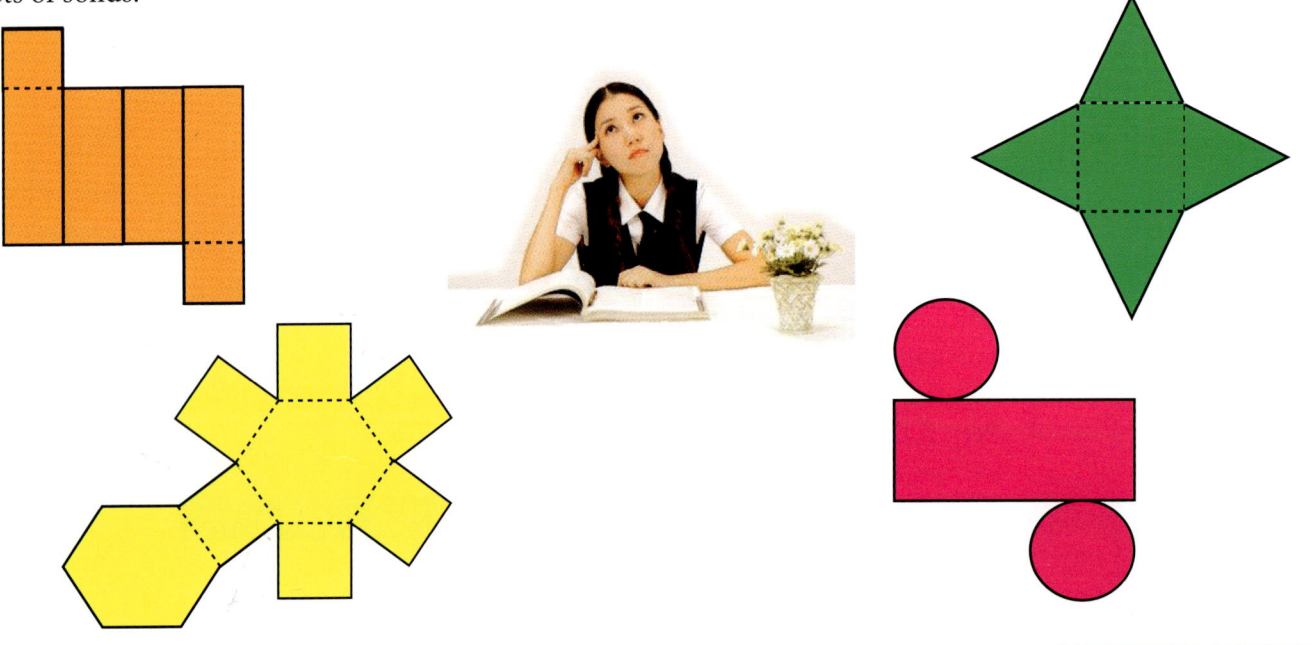

EXPLORE! NETTING A SOLID

Locate a prism to use in this activity.

Step 1: Trace the base of your solid near the edge of a large piece of paper.

Step 2: Tip your solid down on its side. Line up the solid so that the edge of the lateral face touches the corresponding edge on the drawing. Trace this lateral face.

Step 3: Roll the prism on the piece of paper and trace each lateral side, one at a time. Finally, stand the prism up and trace the remaining base. You should never lift the prism.

Step 4: Label each face as top, bottom or side. There will be more than one side. This is a net of your prism.

Step 5: Cut out your net. Fold it on the lines and tape it together. Does it match the figure you started with?

Step 6: Switch prisms with a classmate. Draw a net of your new prism. Will your net be exactly the same as your classmate's net for the same object? Explain how the nets could be different yet still form the same solid.

Step 7: Look at the nets at the beginning of the lesson. What solid will each net form?

EXAMPLE 1 Sketch a net for a triangular prism.

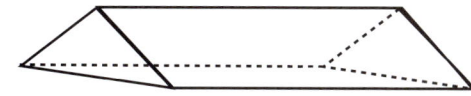

SOLUTION

Step 1: First sketch one base.

Step 2: Sketch one lateral face that connects to the base.

Step 3: Sketch the other base.

Step 4: Finally, sketch the remaining lateral faces.

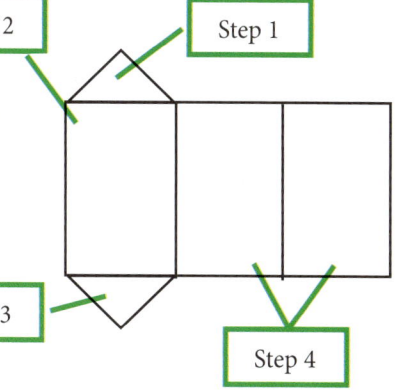

EXAMPLE 2 Draw a net for a pentagonal pyramid.

SOLUTION Start by sketching the base of the pyramid (a pentagon).

Sketch each lateral face (triangle) by attaching it to one of the base edges.

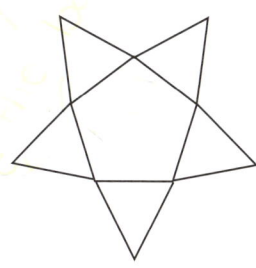

Drawing solids is a skill that requires attention to detail. Solids are three-dimensional. This means some edges in a solid may be hidden. Dashed segments are used for hidden edges.

Drawing a Prism

1. Draw one base.

2. Draw the second base directly above or below the first base.

3. Draw the edges using solid or dashed lines depending on whether or not each edge is hidden.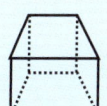

Drawing a Pyramid

1. Draw the base.

2. Draw a point centered above the base.

3. Connect the vertices of the square to the point. Determine whether each edge should be solid or dashed.

Lesson 3.2 ~ Drawing Solids

When drawing the side view of a cylinder or cone use an oval rather than a circle for the base. Draw one or two ovals for the bases of a solid. Connect the 'ends' of each oval to the other base or a vertex. Change any solid lines to dashed lines if they are hidden.

CYLINDER 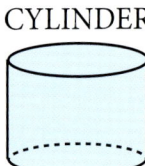   CONE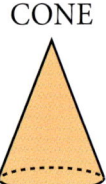

EXERCISES

Sketch a net of each solid.

1.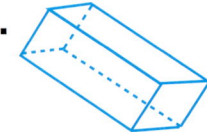

2.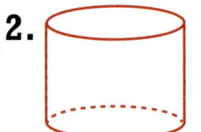

3.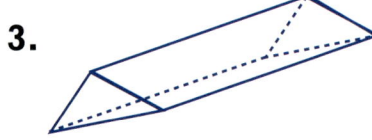

4.

5.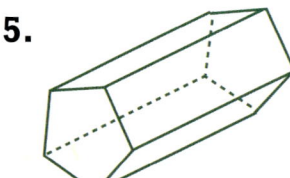

6.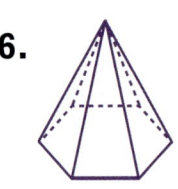

7.

8.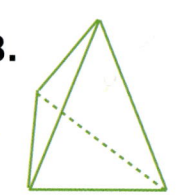

9.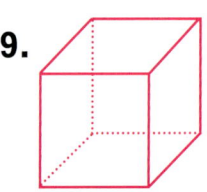

Sketch a diagram of each solid.

10. triangular prism
11. square prism
12. pentagonal prism

13. triangular pyramid
14. square pyramid
15. pentagonal pyramid

16. cone
17. cylinder
18. trapezoidal prism

19. Piper used the net drawn at right to find the number of vertices, edges, lateral faces and bases of a square pyramid. She concluded a square pyramid has 8 vertices, 12 edges, 4 lateral faces and 1 base. Piper is not fully correct. Identify her mistakes and determine the correct number of vertices, edges, lateral faces and bases in a square pyramid.

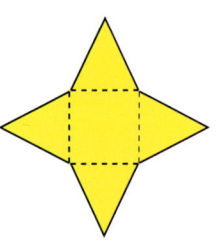

20. Sketch a rectangular prism. Draw two different nets for the rectangular prism.

REVIEW

Identify the number of vertices, edges, lateral faces and bases in each solid.

21.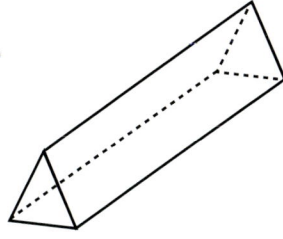

22.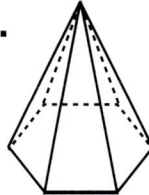

23.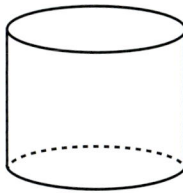

Find the area of each composite figure.

24.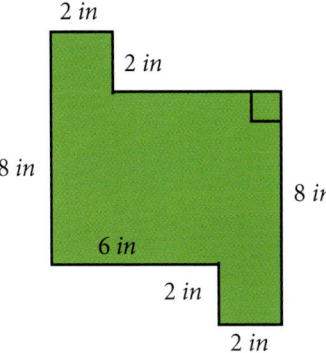

25.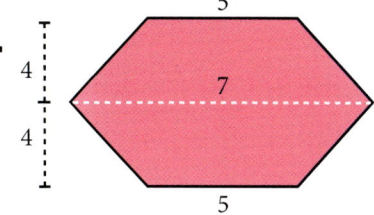

Tic-Tac-Toe ~ Netting a Cube

A cube is a specific type of rectangular prism. The faces of a cube are all congruent squares. Create a minimum of 10 different nets that form the same cube.

Step 1: Draw each net.
Step 2: Cut out each net.
Step 3: Fold to make sure it forms a cube.
Step 4: Present all nets with a creative, visual display.

Tic-Tac-Toe ~ Research Says...

There are many landmarks and buildings that are solids. For example, the Pentagon in Washington, DC is a pentagonal prism with the center removed. The Pyramid du Louvre is a glass and metal pyramid that is an entrance to a museum in Paris, France. Research a famous landmark or building that is a solid. Write a one page paper about the importance of the landmark. On a separate sheet of paper, sketch a diagram of the landmark and calculate the surface area.

Lesson 3.2 ~ Drawing Solids

SLICING SOLIDS

LESSON 3.3

 Describe two-dimensional figures that result from slicing three-dimensional figures.

Molly was playing with a stacking toy of wooden rings that looked similar to a cone when put together correctly. Delaney, her older sister, noticed the top and bottom of each ring was a circle just like the base of the cone. She wondered about slicing any solid parallel to its base. Would the flat surface created by the cut always be the same shape as the base of the original solid?

CUTTING CLAY

EXPLORE!

Step 1: Sketch a cone, a cylinder, a rectangular prism and a square pyramid.

Step 2: What shape is the base of each solid?

Step 3: Analyze each shape and determine what two-dimensional shape would be formed by a cut parallel to the base.

Step 4: Organize this information into the first two columns of a table like the one shown below. The last two columns will be filled in later.

Name of Solid	Shape of Original Base	Shape of New Base after Parallel Cut	Shape of a Side of the Solid	Shape after Making a Perpendicular Cut
Cone				
Cylinder				
Rectangular Prism				
Square Pyramid				

Step 5: Look at the table. What do you notice about each new base formed by a parallel cut compared to the original base?

Step 6: Fill in the column titled "Shape of a Side of the Solid" in the table. Write what shape each side of the solid is. If it does not have a shape, write "none."

Step 7: Look at each solid and determine what two-dimensional shape would be formed if a cut is made perpendicular to the base. Write your answers in the last column of your table.

Step 8: What do you notice about the shape formed by:
      a. making a perpendicular cut to the base in a cone and a pyramid?
      b. making a perpendicular cut to the base in a rectangular prism and a cylinder?

To slice means to cut, making a piece separate from the original shape. In the **Explore!** you determined which two-dimensional shape was formed by different slices. Each slice was either perpendicular to the base or parallel to the base. The table below summarizes the slices made in the **Explore!**.

SLICING SOLIDS		
Name of Solid	Two-dimensional shape formed by a slice <u>parallel</u> to the base	Two-dimensional shape formed by a slice <u>perpendicular</u> to the base
Prism	Shape congruent to its base	Rectangle
Cylinder	Congruent circle	Rectangle
Pyramid	Shape similar to its base	Triangle
Cone	Similar circle	Triangle

Delaney knew that when her mother would slice an orange in half the two-dimensional shape formed along the cut was a circle. She asked her father if any cut on a sphere would make a circle. They took an orange and made a variety of slices. What do you think they found?

SLICING SPHERES	
Sphere	When a sphere is sliced, the slice will always be a circle.

EXAMPLE 1

Determine the two-dimensional shape formed by each slice.
a. A slice perpendicular to the base of a cone.
b. A slice parallel to the base of a triangular prism.

SOLUTIONS

a. A <u>triangle</u> is formed by slicing a cone perpendicular to the base.

b. A <u>triangle</u> is formed by slicing a triangular prism parallel to its base.

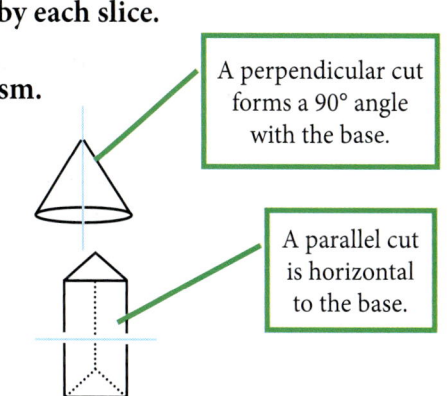

A perpendicular cut forms a 90° angle with the base.

A parallel cut is horizontal to the base.

Lesson 3.3 ~ Slicing Solids

EXAMPLE 2   A cut was made on each solid parallel to its base. Determine if the two-dimensional figure formed by the cut is similar or congruent to its base.

a. 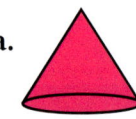   b. 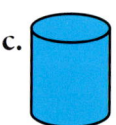   c. (cylinder)

Solutions

a. <u>SIMILAR</u>. The cut formed will be a circle but it will be smaller than the base.

b. <u>CONGRUENT</u>. The cut formed will be a hexagon the same size as the base.

c. <u>CONGRUENT</u>. The cut formed will be a circle that is the same size as the base.

A solid can be sliced many ways. Each cut will not always be parallel or perpendicular to the solid's base. The diagram at the right shows the corner of a cube being sliced. The two-dimensional shape made by the cut is a triangle.

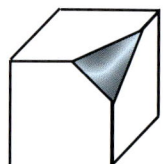

EXERCISES

Use the solids below. Read each description. Using the word bank, list the two-dimensional shape formed by each slice.

      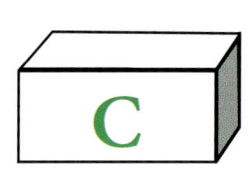   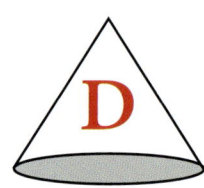   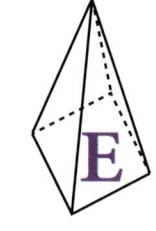

Cylinder   Cube   Rectangular Prism   Cone   Square Pyramid

1. Figure B sliced perpendicular to the base

2. Figure D sliced parallel to the base

3. Figure A sliced parallel to the base

4. Figure A sliced perpendicular to the base

5. Figure E sliced perpendicular to the base

6. Figure D sliced perpendicular to the base

7. Figure C sliced perpendicular to the base

8. Figure B sliced diagonally from top left to bottom right

9. Figure C sliced parallel to the base

10. Figure E sliced parallel to the base

Word Bank:
- CIRCLE
- OVAL
- PARALLELOGRAM
- RECTANGLE
- SQUARE
- TRIANGLE

11. Pedro cut a cube parallel to its base.
   a. What shape is the base of the new solid?
   b. The two new solids that are formed are no longer cubes. What are the names of the two new solids?

12. Cheri cut a cube diagonally from the top right edge to the bottom left edge. What shape is formed by the cut? What other solid, when cut diagonally, will have the same shape as the diagonal cut of the cube?

13. A cylinder is 12 inches tall and has a diameter of 8 inches.
   a. A cut is made perpendicular to the base of the cylinder and through the center of the cylinder. What is the area of the surface created with the perpendicular cut? Use mathematics to justify your answer.
   b. A cut is made parallel to the base of the cylinder and through the center of the cylinder. What is the area of the surface created with the parallel cut? Use mathematics to justify your answer.

14. A rectangular prism has a length of 4 *in*, a width of 2 *in* and a height of 6 *in*. Judy says the rectangle created by a cut parallel to the base has an area of 8 *in*². Ashiraf says the rectangle created by a cut parallel to the base has an area of 12 *in*². How can they both be correct? Explain your reasoning.

Each solid is cut parallel to its base. Determine the name of the two-dimensional figure formed by the cut and if the figure is similar or congruent to its base.

15. 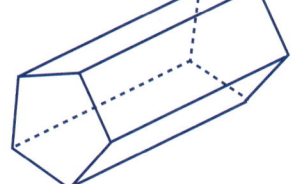   **16.**    **17.**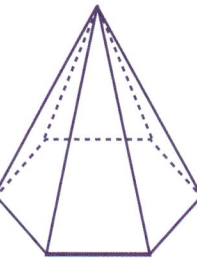

18. Nathan sliced a solid once. The cut was parallel to the solid's base. The figure formed was a pentagon congruent to the solid's base. Name the original solid.

19. Sylvette made one slice on two different solids. She cut both solids perpendicular to their bases. The two-dimensional figure formed by each cut was a triangle. Name the two solids that Sylvette sliced.

20. Sketch a composite solid that when cut parallel to its base is sometimes congruent to its base and sometimes similar to its base. In your sketch, identify a cut that creates a similar figure to the base and a cut that creates a congruent figure to the base.

Name the two-dimensional shape formed by each slice.

21. 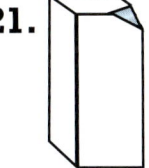   **22.** 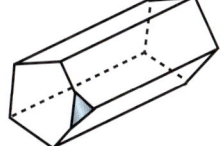   **23.** 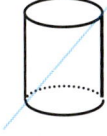   **24.**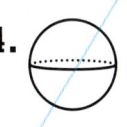

REVIEW

Draw a net for each solid.

25.

26.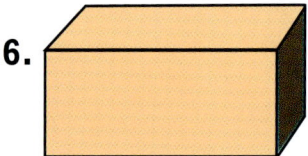

27.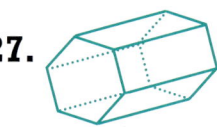

Find each missing measure. Use 3.14 for π. Round to the nearest hundredth, as needed.

28. Circumference ≈ 81.4 m

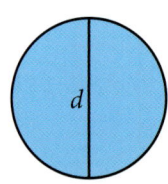

29. Area ≈ 84 in^2

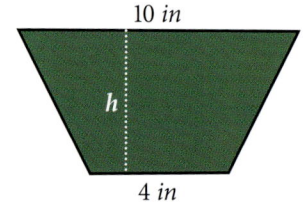

30. Area = 65.25 ft^2

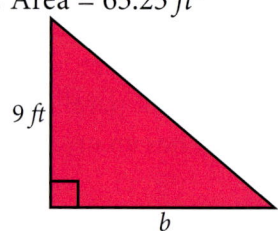

Tic-Tac-Toe ~ A Solid View

There are five three-dimensional figures in the table below. Complete the table by sketching the left side view, right side view, top view, bottom view and rear view of each solid.

	Left Side View	Right Side View	Top View	Bottom View	Rear View

Lesson 3.3 ~ Slicing Solids

SURFACE AREA OF PRISMS

LESSON 3.4

 Calculate the surface area of prisms.

The **surface area** of a solid is the sum of the areas of all the surfaces. This includes the areas of the bases and faces of the solid. The surface area of a prism is the sum of the area of the two bases and the area of the lateral faces. A lateral face is a side of a solid that is not a base.

EXPLORE!   TAKE YOUR PICK

Step 1: Sketch the net of the rectangular prism at the right.

Step 2: Label the edges of each face of the net with the correct lengths.

Step 3: Are there any rectangles that are the same shape and size? If so, how many are there of each size?

Step 4: Calculate the area of each face of the prism.

Step 5: Find the sum of the face areas. This sum is the surface area of the prism.

Step 6: Surface area can also be determined without drawing a net. Find the area of the bottom base of the prism used in **Step 1**.

Step 7: What is the area of the top of the prism?

Step 8: Find the perimeter of the base. Multiply the perimeter by the height of the prism.

Step 9: Add the answers from **Steps 6–8**. This sum is the surface area of the prism.

Step 10: What do you notice about the answers in **Step 9** and **Step 5**? Discuss what you observed with a classmate.

Step 11: Another formula that can be used to find the surface area of a rectangular prism is: $2(lw + wh + lh)$.
    a. Use this formula to find the surface area of the prism above.
    b. Explain why this formula works.

Step 12: Find the surface area of each prism. Try at least two different methods from this **Explore!**.

    a.
    b.
    c.

The lateral surface area is the sum of the areas of the lateral faces. The total surface area includes the area of the two bases.

LATERAL SURFACE AREA OF A PRISM

The lateral area (LA) of a prism is equal to the perimeter (P) of the base times the height (h) of the prism.

LA = Sum of Lateral Faces
LA = Ph

SURFACE AREA OF A PRISM

The surface area of a prism is equal to the sum of the lateral area (LA) and the area of the two bases (B).

SA = Sum of all Faces
SA = LA + 2B

EXAMPLE 1 The dimensions of a shipping container are given. Find the surface area of the container.

4 ft, 2 ft, 3 ft

SOLUTION

Find the perimeter of the base.   4 + 2 + 4 + 2 = 12
Locate the height of the prism.   h = 3
Find the area of the base.   4(2) = 8   LA = Ph

Use the surface area formula for a prism.   SA = LA + 2B
Substitute all known values for the variables. SA = 12(3) + 2(8)
Multiply, then add.   SA = 36 + 16 = 52

The surface area of the shipping crate is 52 square feet.

EXAMPLE 2 The surface area of a pentagonal prism is 1,445 square centimeters. The lateral area of the prism is 1,100 square centimeters. What is the area of one base?

SOLUTION

Use the basic surface area formula for a prism.   SA = LA + 2B
Substitute all known values for the variables.   1445 = 1100 + 2B
Solve for B.   −1100 −1100
Divide both sides by 2. $\frac{345}{2} = \frac{2B}{2}$

                                                     172.5 = B

B represents the area of one base of the prism.

The area of one base in the pentagonal prism is 172.5 cm^2.

102 Lesson 3.4 ~ Surface Area of Prisms

EXERCISES

1.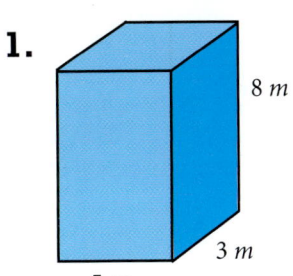
a. Draw a net of the rectangular prism at the left. Label the lengths of each side.
b. Find and write the area of each polygon in the net.
c. Add all areas to find the surface area of the rectangular prism.

2.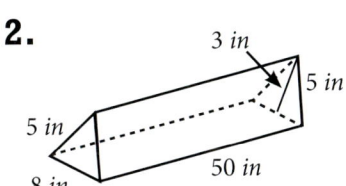
a. Draw a net of the triangular prism at the left. Label the key information on each polygon.
b. Find and write the area of each polygon in the net.
c. Add all areas to find the surface area of the triangular prism.

Find the surface area of each prism.

3.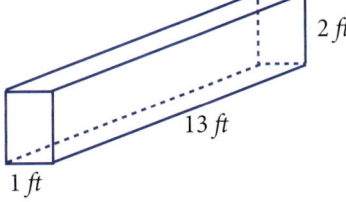

4.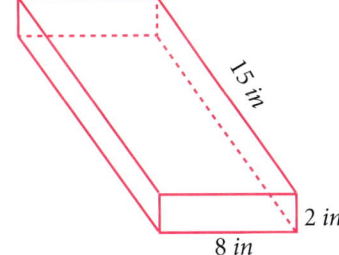

5.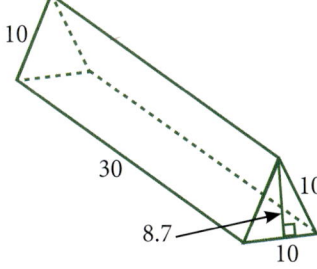

6.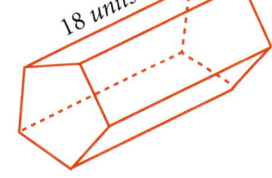
Base area = 60 u^2
Perimeter of base = 30 units

7.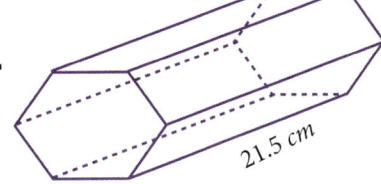
Area of the base = 184 cm^2
Each side of the base is 8.4 cm

8.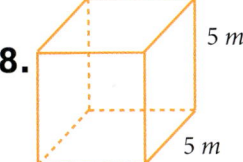

9. The area of one base on a shoe box is 120 in^2. The lateral area is 220 in^2. Find the surface area of the shoe box. Zuleyma says the surface area of the shoe box is 120 + 220 = 340 in^2. She has made a mistake. Identify Zuleyma's mistake and find the surface area of the shoe box.

10. The lateral area of an octagonal prism is 192 square meters. The area of one base is 43 square meters. What is the surface area of the octagonal prism?

11. The area of one base of a triangular prism is 18 square feet. The perimeter of the triangular base is 19 feet. The height of the prism is 21 feet. Find the surface area of the triangular prism.

Lesson 3.4 ~ Surface Area of Prisms 103

12. An octagon with an area of 120 cm^2 is the base of a prism. Each side of the octagon is 5 cm. The height of the prism is 11 cm. Find the surface area of the octagonal prism.

13. The surface area of a triangular prism is 72 ft^2. The lateral area is 60 ft^2. Find the area of one base. Use mathematics to justify your answer.

14. A cargo box has these dimensions: 5 m by 6 m by 3.2 m. Find the surface area of the box using the formula: $2(lw + wh + lh)$.

15. Sergio wrapped a gift for his mother. The box was a rectangular prism with dimensions of 12 inches by 10 inches by 4 inches.
  a. How much wrapping paper did Sergio need to exactly cover the box?
  b. Sergio ended up using 450 square inches of wrapping paper. Why do you think he used more than the answer in **part a**? Explain your reasoning.

16. The surface area of a rectangular prism is 120 square inches. The lateral area is 90 square inches.
  a. What is the area of one base?
  b. Identify a possible set of dimensions for the base of the prism.
  c. Find the height of the prism.
  d. Draw a diagram of the prism. Label the length, width and height.

17. A rectangular prism has a surface area of 846 square feet. Its height is 12 feet and its width is 9 feet. Use the formula $2(lw + wh + lh)$ to find the length of the prism. Show all work necessary to justify your answer.

REVIEW

Find the area of the shaded regions.

18.

19.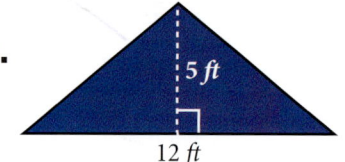

20.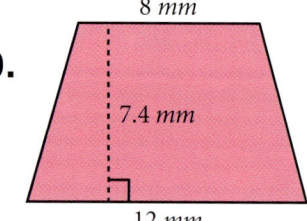

21.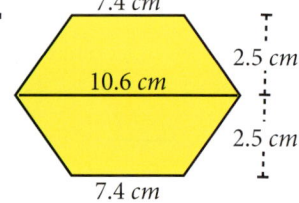

22.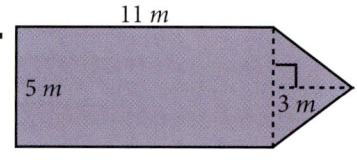

23. Use $\frac{22}{7}$ for π.
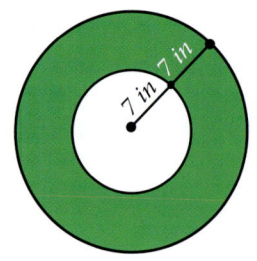

Tic-Tac-Toe ~ Stencils

Create a set of net stencils a student could use to make their own solids. The set of nets should include:
- two different rectangular prisms
- a triangular prism
- a pentagonal or hexagonal prism
- a square pyramid
- a pentagonal or hexagonal pyramid
- two different cylinders

For a final presentation, include the stencils and a sample set of solids made from the stencils.

Tic-Tac-Toe ~ Volume Of A Sphere

The volume of a sphere can be found using the formula $V = \frac{4}{3}\pi r^3$.

Example:

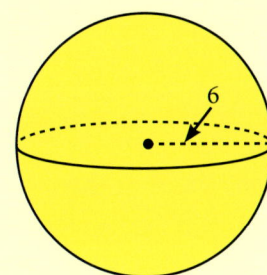

$V = \frac{4}{3}\pi r^3$
$\approx \frac{4}{3}(3.14)(6)^3$
≈ 904.32
$V \approx 904.32$ cubic inches

Find the volume of each sphere. Use 3.14 for π. Round to the nearest hundredth.

1. 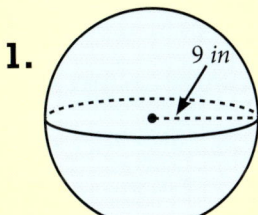 (9 in)

2. 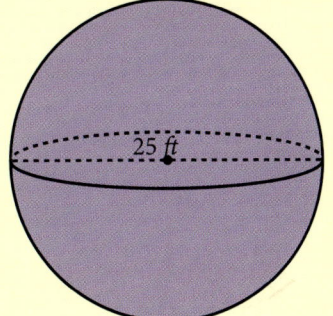 (25 ft)

3. 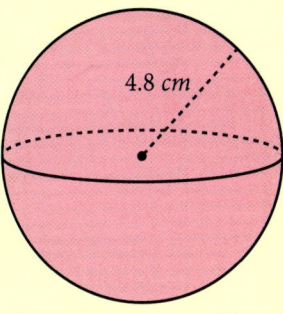 (4.8 cm)

4. $r = 2$ inches

5. $d = 7$ feet

6. $d = 12$ meters

7. Find the volume of at least two balls (basketball, golf ball, softball, etc). All measurements should be in centimeters. Use 3.14 for π.
   a. When a sphere is sliced through its center, a great circle is formed. Without slicing each ball, use a piece of string to measure the circumference of a great circle on each ball.
   b. Find the radius of each ball.
   c. Calculate the volume of each ball.
   d. Describe what material or substance makes up the volume of each ball. This may require some research.

VOLUME OF PRISMS

LESSON 3.5

 Calculate the volume of prisms.

Mischa needs laundry detergent. Her favorite detergent comes in two different types of containers. One container is a rectangular prism and the other is a triangular prism. They cost the same amount but she wants to buy the container that holds the most detergent. The container with the most detergent has the most volume.

Volume is the number of cubic units needed to fill a solid. Examples of these units are cubic inches (in^3) or cubic centimeters (cm^3). Think of volume as the number of cubes measuring one unit on each side (*length*, *width* and *height*) that fit inside of a prism.

EXPLORE!

CUTTING CORNERS

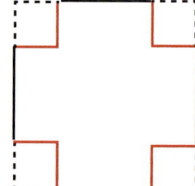

Step 1: Cut out a square that is 4 inches on each side.

Step 2: Cut a one-inch square out of each corner of the square. Fold each side up to form a square prism and tape the edges together.

Step 3: What are the dimensions of the prism? Identify the length, width and height.

Step 4: How many one-inch cubes do you think would fit in the prism? Explain how you arrived at this answer.

Step 5: Find the area of the base of the folded prism. Multiply the area of the base times the height. This is the number of one-inch cubes that will fit in the prism. How does the number of cubes that fit in the prism compare to your estimate?

Step 6: Write a formula to help you calculate the volume of a prism. Use *B* for the area of the base and *h* for the height.

Step 7: Find the volume of the following prisms using the formula from **Step 6**.

a.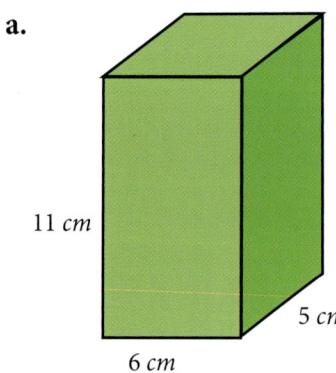

b.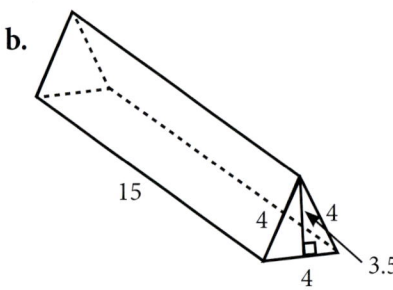

c.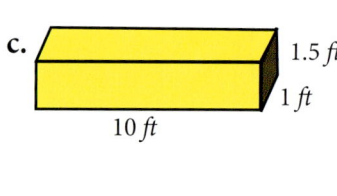

The volume of a prism is the number of cubic units needed to fill the prism. The figure on the right shows a prism with a volume of 18 cubic units. The volume of a prism is found by multiplying the area of the base times the height.

VOLUME OF A PRISM

The volume of a prism is equal to the product of the area of the base (*B*) and the height (*h*).

Volume = (area of the base)(height)

$V = Bh$

It is important to correctly determine which faces of a prism are the bases to calculate the volume. The bases are the faces that are parallel to one another and congruent.

EXAMPLE 1 Find the volume of each prism.

a.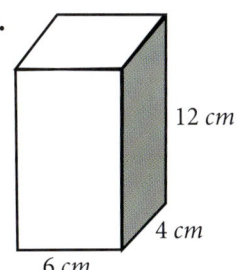
b. Base area = 41 *cm²*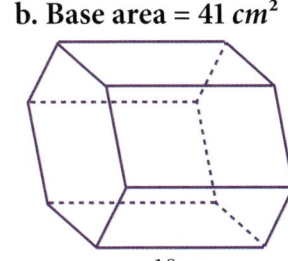
c.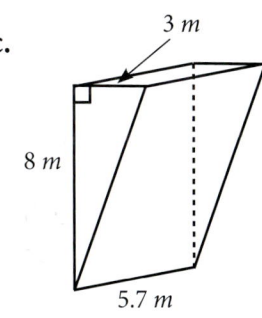

SOLUTIONS Volume of a prism = (area of the base)(height) = *Bh*

The bases are triangles.

a. $V = Bh$
 $= (6 \cdot 4)(12)$
 $= (24)(12)$
 $= 288$

b. $V = Bh$
 $= (41)(10)$
 $= 410$

The bases are hexagons.

c. $V = Bh$
 $= (\frac{1}{2} \cdot 8 \cdot 3)(5.7)$
 $= (12)(5.7)$
 $= 68.4$

Volume = 288 *cm³*   Volume = 410 *cm³*   Volume = 68.4 *m³*

EXAMPLE 2 Mischa can buy her favorite laundry detergent in two different-shaped containers. One container is a triangular prism with a base area of 15 square inches and a height of 6 inches. She can also buy the detergent in a rectangular prism container with a length of 5 inches, a width of 4 inches and a height of 4.5 inches. Which container has the most volume?

SOLUTION

Triangular box
$V = Bh$
 $= (15)(6)$
 $= 90$

Rectangular box
$V = Bh$
 $= (lw)(h)$
 $= (5 \cdot 4)(4.5)$
 $= (20)(4.5)$
 $= 90$

Both boxes have 90 cubic inches of volume. They are equal in volume.

Lesson 3.5 ~ Volume of Prisms

EXERCISES

Find the volume of each prism.

1.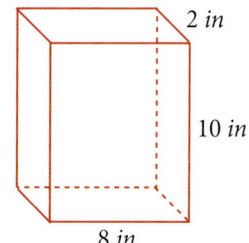

2.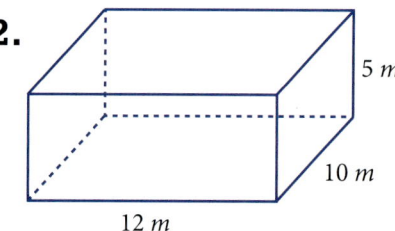

3. Base = 21 cm^2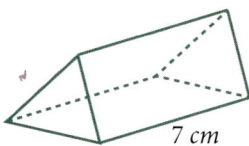

4. A hexagonal sandbox has a base area of 24 square feet. The sides of the sandbox are 1 foot high.
   a. How much sand does it take to fill the sandbox?
   b. Willow fills the sandbox with sand until the height of the sand is $\frac{1}{4}$ foot from the top of the sandbox. Find the volume of the sand in the sandbox. Show all work necessary to justify your answer.

5. A rectangular prism is 5 feet long, 3 feet wide and 9 feet tall.
   a. Find the volume of the rectangular prism.
   b. The rectangular prism is cut into two equally-sized triangular prisms. Calculate the volume of each triangular prism.
   c. What is the relationship between the volume of the rectangular prism and one of the triangular prisms?

6. Some resources show the formula for a rectangular prism as: V = lwh. Use this formula to calculate the volume of the following rectangular prisms.

   a.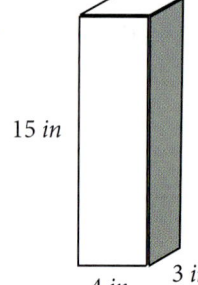

   b.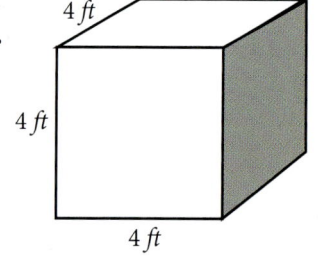

   c.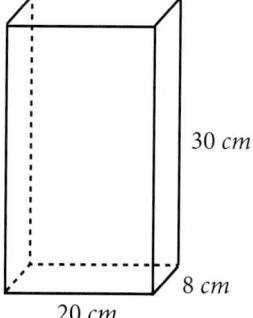

   d. Explain how the formulas, V = lwh and V = Bh are equivalent for rectangular prisms.

7. Janis measured the sides of the box shown at right. The height of the box is 3 inches. The width is 7 inches. The length of the box is 10 inches. She decided to fill the box with pieces of fudge that are one cubic inch each.
   a. Find the volume of the box.
   b. How many pieces of fudge fit in the box?
   c. Janis decreased the size of the fudge to cubes that are 0.5 inches on each side. How many pieces of this size of fudge fit in the box? Use words and/or numbers to show how you determined your answer.

108 Lesson 3.5 ~ Volume of Prisms

8. A fish aquarium has a base that is 200 *cm* long and 40 *cm* wide. The aquarium is 50 *cm* tall.
   a. Find the volume of the fish aquarium.
   b. Each fish requires 3,800 cubic centimeters of water. What is the maximum number of fish the tank should hold? Explain your reasoning.

Find the volume of each prism.

9. Base area = 23 *in²*

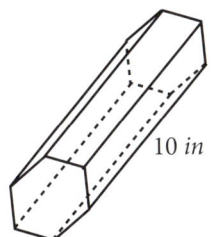

10.

6 in, 18 in, 9 in

11.

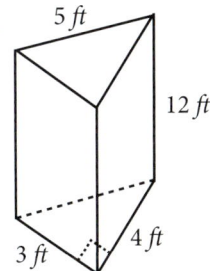

5 ft, 12 ft, 3 ft, 4 ft

12. The top of a bolt is the shape of a hexagon. The area of the head of the bolt is 43.3 *mm²*. The bolt is 3.5 *mm* tall. What is the volume of the bolt head?

13. An 8-inch tall octagonal water glass has a base area of $4\frac{7}{8}$ *in²*. Find the volume of the glass.

14. Pasta is available in a 12-inch tall container shaped like a triangular prism. The base of the prism is 3.75 square inches. Pasta is also available in a rectangular prism with a base that is 2 inches by $\frac{3}{4}$ inch. This box is also 12 inches tall. Both containers are the same price. Which container gives you more pasta for your money? Show all work necessary to justify your answer.

15. A sink is 1 foot deep, 3 feet long and 1.5 feet wide.
   a. What is the volume of the sink in cubic feet?
   b. What is the volume of the sink in cubic inches?
   c. Which volume would most likely be used to describe the amount of water that a sink can hold? Explain your reasoning.

16. A rectangular prism has dimensions of 4 *in* by 5 *in* by 6 *in*. To find the volume of the prism, Sharika says it does not matter if she uses the 4 *in* by 5 *in*, the 5 *in* by 6 *in* or the 4 *in* by 6 *in* face as the base. Do you agree or disagree? Use words and/or numbers to show how you determined your answer.

17. The volume of a rectangular prism is 36 *in³*. Give two different sets of possible measurements for its length, width and height.

18. Below are the nets of two prisms. Find the volume of each prism.

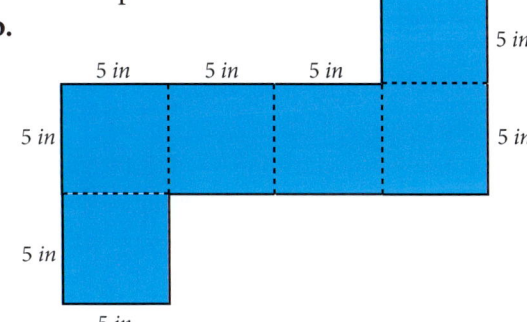

Lesson 3.5 ~ Volume of Prisms

REVIEW

Find the circumference of each circle. Use $\frac{22}{7}$ for π.

19.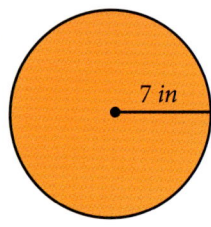
7 in

20.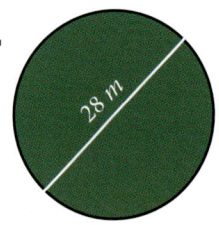
28 m

21.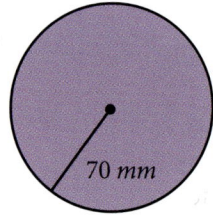
70 mm

22. A circular flower bed has a diameter of 20 feet. What is the perimeter of the flower bed? Use 3.14 for π.

23. A cylindrical gift box has a radius of 8 inches and a height of 20 inches. How much wrapping paper is needed to cover the box with no overlap? Use 3.14 for π.

Tic-Tac-Toe ~ More Cereal Please

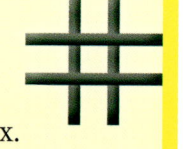

Breakfast cereal comes in a variety of box sizes. In this activity you will compare the volume of a box to the weight of the cereal in the box.

Step 1: Locate at least 6 different types of cereal boxes at home or at the store. Measure the length, width and height of each box to the nearest quarter of an inch. Record the measurements in the table below.

Name of Cereal	Length	Width	Height	Volume	Ounces

Step 2: Complete the table by calculating the volume of each box in cubic inches. Record the weight of the cereal in ounces as shown on the cereal box.

Step 3: Find the unit rate (ounces per cubic inch) for each box of cereal. Display your findings in an organized manner below the table.

Step 4: Which type of cereal(s) had the highest amount of ounces per cubic inch? Which type of cereal(s) had the lowest amount of ounces per cubic inch?

Step 5: Do you notice any trends in types of cereal and their rate of ounces per cubic inch? If so, explain.

Step 6: Look up the word 'density.' Record the definition. How does this word relate to this activity? Write a sentence that describes one of your findings in this activity. Your sentence must include the word 'density.'

SURFACE AREA OF REGULAR PYRAMIDS

LESSON 3.6

 Calculate the surface area of regular pyramids.

A pyramid is a solid with a polygonal base and triangular lateral faces that meet at a vertex. In this lesson, you will work with regular pyramids. The base of a regular pyramid is a polygon with sides of equal length and angles of equal measure. The **slant height** of a pyramid is the height of a lateral face. The variable l is used to represent slant height. The net of a square pyramid is shown below.

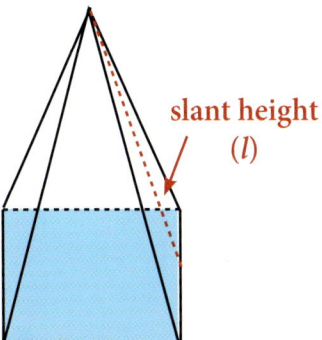

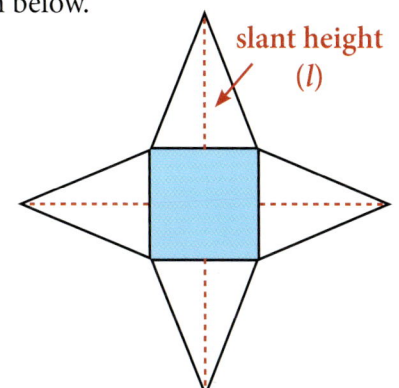

EXPLORE!

TENT MAKING

Tyrisha's family is planning a barbecue on the beach over Spring Break. They want to make a tent out of a tarp in case it rains. Tyrisha decides the tent should be an enclosed square pyramid.

Step 1: Draw a two-dimensional pattern (net) for Tyrisha's tent. The sides of the base will be 10 feet. The slant height of the tent will be 8 feet.

Step 2: Find the area of the base of the tent.

Step 3: Find the area of each lateral face of the tent.

Step 4: How much tarp is needed to exactly cover the lateral faces of the tent?

Step 5: Tyrisha's brother, Tyrone, says he would not need a net to find the lateral area. He says, "The lateral area of the pyramid is one-half of the perimeter of the base times the slant height." Check his statement. Do you agree or disagree with Tyrone?

Step 6: Tyrone's formula only helps calculate the lateral area. They plan to enclose the entire space by putting tarp down for the base of the tent. Write a formula that will help you find the entire surface area of the tent.

Step 7: Tyrisha decides her tent should be a bit larger. The sides of the base of the larger tent will be 12 feet. The slant height of the tent will be 9 feet. Find the total amount of tarp needed. Show all work necessary to justify your answer.

Lesson 3.6 ~ Surface Area of Regular Pyramids

LATERAL SURFACE AREA OF A REGULAR PYRAMID	SURFACE AREA OF A REGULAR PYRAMID
The lateral area (LA) of a pyramid is equal to half the perimeter (P) of the base times the slant height (*l*) of the pyramid. LA = Sum of Lateral Faces LA = $\frac{1}{2}Pl$	The surface area of a pyramid is equal to the sum of the lateral area (LA) and the area of the base. SA = Sum of all Faces SA = LA + B SA = $\frac{1}{2}Pl + B$

EXAMPLE 1 Find the surface area of the regular pentagonal pyramid given that the base area is 139 square centimeters.

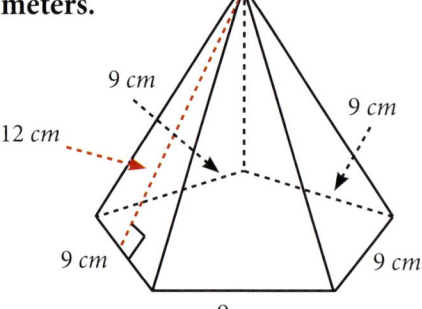

SOLUTION

Find the perimeter of the base.	9(5) = 45
Identify the slant height of the pyramid.	*l* = 12
Use the surface area formula for a pyramid.	SA = LA + B
Substitute all known values for the variables.	SA = $\frac{1}{2}$(45)(12) + 139
Multiply, then add.	SA = 270 + 139 = 409

LA = $\frac{1}{2}Pl$

The surface area of the pyramid is 409 square centimeters.

EXAMPLE 2 Find the surface area of the square pyramid.

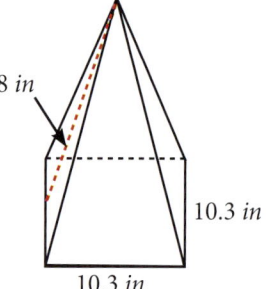

SOLUTION

Find the perimeter of the base.	10.3(4) = 41.2
Find the area of the base.	10.3(10.3) = 106.09
Identify the slant height of the prism.	*l* = 8
Use the surface area formula for a pyramid.	SA = LA + B
Substitute all known values for the variables.	SA = $\frac{1}{2}$(41.2)(8) + 106.09
Multiply, then add.	SA = 164.8 + 106.09 = 270.89

LA = $\frac{1}{2}Pl$

The surface area of the pyramid is 270.89 square inches.

EXERCISES

Determine the number of lateral faces on each pyramid.

1. octagonal pyramid
2. square pyramid
3. hexagonal pyramid
4. heptagonal pyramid
5. triangular pyramid
6. pentagonal pyramid

7.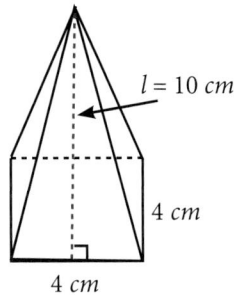
   a. Draw a net of the pyramid.
   b. Find the area of each polygon in the net.
   c. Find the surface area of the pyramid.

8.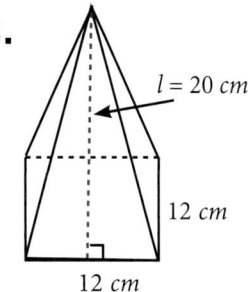
   a. Draw a net of the pyramid.
   b. Find the area of each polygon in the net.
   c. Find the surface area of the pyramid.

9. The base of a regular pentagonal pyramid has a perimeter of 60 feet. The slant height of the pyramid is 9 feet. Find the lateral area of the pyramid.

10. A square pyramid has a base edge that measures 8 meters and a slant height of 30 meters. Marcus says the lateral area of the square pyramid is $\frac{1}{2}(8)(30) = 120$ square meters. Marcus has an error in his work. Identify the error and find the lateral area of the square pyramid.

Find the lateral area of each pyramid.

11. Slant height = 11
    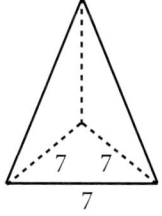

12.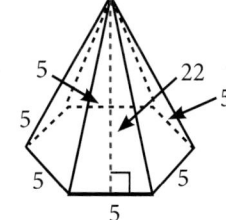

Find the surface area of each pyramid.

13.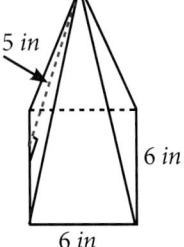

14.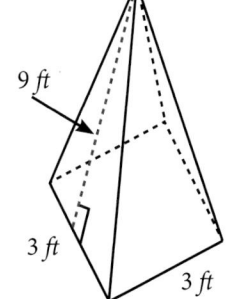

15. A regular triangular pyramid has a slant height of 10 inches. The perimeter of the base is 24 inches. The base of the pyramid has an area of 27.7 square inches. Find the surface area of the pyramid.

16. Terry made game pieces in the shape of square pyramids. Each piece has a base edge of 2 cm and a slant height of 4 cm. He will paint all of the pieces. He wants to know how much paint he needs.

   a. Find the surface area of one game piece.
   b. Each game has 24 game pieces. Find the total surface area of one set of game pieces.
   c. He wants to make 12 games. What is the total surface area for all 12 games?
   d. A can of paint covers 400 square centimeters. How many cans of paint will he need? Use words and/or numbers to show how you determined your answer.

17. A regular hexagonal pyramid has a base area of 392.9 square feet. The sides of the hexagon are 12.3 feet long. The slant height of the pyramid is 15.9 feet. What is the surface area of the pyramid?

18. A square pyramid has a perimeter of 50 inches and a slant height of $15\frac{3}{4}$ inches. Find the surface area of the pyramid. Show all work necessary to justify your answer.

19. A square pyramid has a base area of 64 cm^2. The slant height of the pyramid is 7 cm. What is the surface area of the pyramid?

20. The surface area of a square pyramid is 220 square inches. Each edge of the base has a length of 10 inches. What is the slant height of the square pyramid? Use mathematics to justify your answer.

21. The surface area of a regular hexagonal pyramid is 162 square feet. The perimeter of the base is 24 feet and the slant height is 10 feet. What is the area of the base? Use mathematics to justify your answer.

REVIEW

Calculate the surface area of each solid.

22.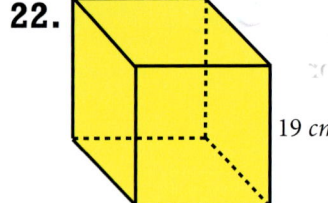

23.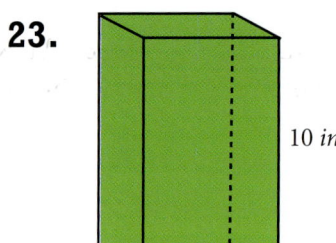

24.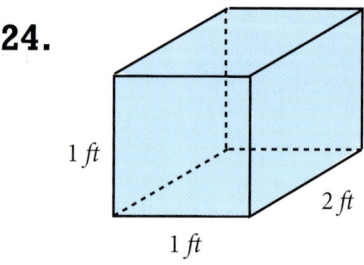

25. How many edges does a hexagonal pyramid have?

26. How many vertices are on a pentagonal prism?

27. How many lateral faces are on an octagonal prism?

VOLUME OF PYRAMIDS

LESSON 3.7

  Calculate the volume of pyramids.

The volume of a pyramid and prism with congruent bases and equal heights have a special relationship. In the **Explore!** you will investigate that relationship. A pyramid has two different types of height. When finding surface area, the slant height (l) is used. When finding the volume, you must know how tall the pyramid stands. This is the regular height (h).

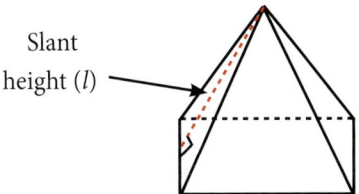
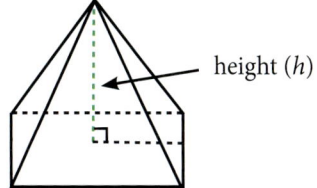

EXPLORE! PYRAMID VS PRISM

Make or find a pyramid and prism that have congruent bases and the same height to use in this activity.

Step 1: Estimate how many times larger the volume of the prism is compared to the pyramid.

Step 2: Fill the pyramid with rice, beans or popcorn kernels. Pour the contents of the pyramid into the prism. Repeat until the prism is full.

Step 3: How many times did you need to empty the pyramid to fill the prism?

Step 4: What fraction is the volume of the pyramid compared to the prism?

Step 5: The volume of a prism is found using the formula $V = Bh$. How could you modify this formula to find the volume of a pyramid? Write the formula for the volume of a pyramid.

Step 6: Use the formula to find the volume of each pyramid.

a.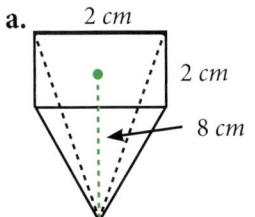

b.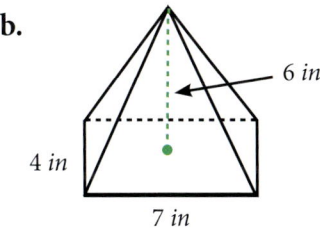

c.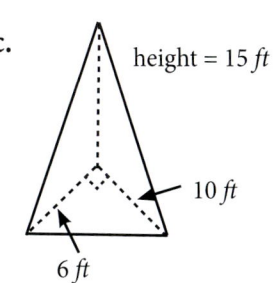

Lesson 3.7 ~ Volume of Pyramids   115

> **VOLUME OF A PYRAMID**
>
> The volume of a pyramid is equal to one-third of the product of the base area (B) and the height (h).
>
> $$V = \tfrac{1}{3}(\text{area of the base})(\text{height})$$
>
> $$V = \tfrac{1}{3}Bh \text{ or } V = \frac{Bh}{3}$$

EXAMPLE 1 — A 15-foot tall pentagonal pyramid has a base area of 38 square feet. Find the volume of the pyramid.

SOLUTION

Use the volume formula for a pyramid.	$V = \tfrac{1}{3}Bh$
Substitute all known values for the variables.	$V = \tfrac{1}{3}(38)(15)$
Multiply.	$V = 190$

The volume of the pyramid is 190 ft^3.

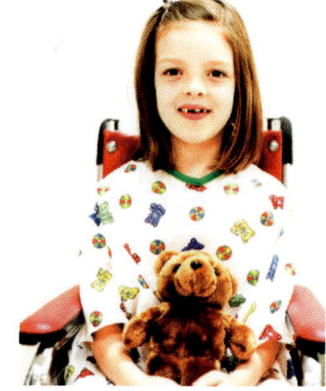

EXAMPLE 2 — Crystal received an award for being an outstanding hospital patient. The award was a square pyramid shape. The base of the pyramid was 8 centimeters on each side. It was 10 centimeters tall. Find the volume of the award. Round to the nearest whole number.

SOLUTION

Find the area of the base.	$B = 8(8) = 64$
Use the volume formula for a pyramid.	$V = \tfrac{1}{3}Bh$
Substitute all known values for the variables.	$V = \tfrac{1}{3}(64)(10)$
Multiply.	$V = 213.\overline{3}$
Round to the nearest whole number.	≈ 213

The volume of Crystal's award was a little more than 213 cubic centimeters.

EXERCISES

1. What is the relationship between the volume of a pyramid and prism which have congruent bases and are the same height?

2. A pyramid and prism are the same height and have congruent bases. The prism is cut in half. Which solid has more volume? Explain your reasoning.

3. The volume of a pyramid is 120 cm^3. What is the volume of a prism that has a congruent base and is the same height as the pyramid?

4. A prism holds 45 gallons of water. How many gallons of water does a pyramid with the same base and height hold? Use words and/or numbers to show how you determined your answer.

5. It takes 1 hour and 30 minutes to fill a rectangular swimming pool. How long will it take to fill a rectangular pyramid that is as deep as the pool and has a base congruent to the pool?

Find the volume of each pyramid. Let B represent the base area and h represent the height.

6. $B = 6\ m^2$, $h = 10\ m$

7. $B = 22\ in^2$, $h = 12\ in$

8. $B = 12.3\ cm^2$, $h = 1.5\ cm$

Find the volume of each pyramid.

9.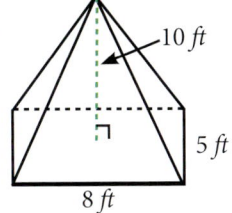

10.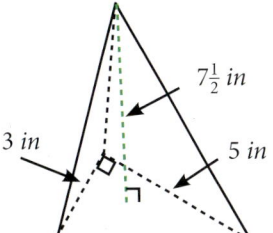

11.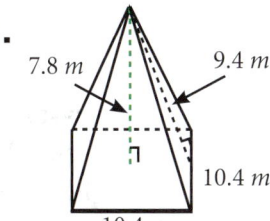

12. Celia makes square pyramidal candles. She buys the wax in the form of rectangular prisms. Each block of wax is 8 inches long, 5 inches wide and 3 inches tall. She melts the wax and pours it into a square pyramidal mold with a base of 2 inches on each side and a height of 4.5 inches.
    a. What is the volume of one candle?
    b. About how many candles can be made from one block of wax? Show all work necessary to justify your answer.

13. Duncan received the spelling bee award for his school district. The superintendent came to the school assembly to give Duncan a trophy. The trophy was made of aluminum and was shaped like a square pyramid. The base of the pyramid had a perimeter of 40 cm. It was 12 cm tall. What was the volume of the trophy?

14. Sally built a sand castle pyramid. The pyramid had a square base that was 24 inches on each side. The sand castle was built with 1,600 cubic inches of sand. Find the height of the sand castle. Use mathematics to justify your answer.

15. A paperweight shaped like a triangular pyramid is $2\frac{1}{2}$ inches tall. The base of the paperweight is a triangle with a base length of 3 inches and a height of 2 inches. Find the volume of the paperweight.

REVIEW

Find the surface area of each solid. Use 3.14 for π.

16.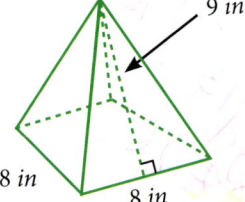

17.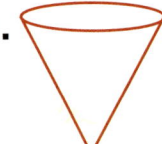

Determine what two-dimensional shape is formed when a cut is made parallel to the base. Then determine if the shape will be similar or congruent.

18.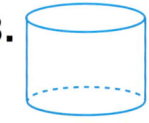

19.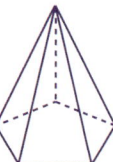

20.

21.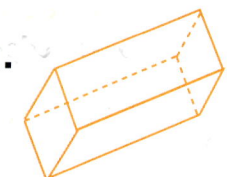

Tic-Tac-Toe ~ A Solid Album

Geometric solids are all around you in the world. Create an album of solids. Include a picture of a solid on each page of your album. The pictures can be photographs, newspaper or magazine clippings or pictures printed off the internet. Each page of the album should also include the following:

- A description of the object (what it is and where it can be found).
- The geometric name of the solid.
- The number of lateral faces, bases, edges and vertices, if applicable.
- Source of the picture (i.e. name of magazine, internet site, etc).

The album should include at least 10 pages with a front and back cover. At least one of each type of solid (prism, pyramid, cylinder, cone and sphere) should be included.

REVIEW

BLOCK 3

Vocabulary

base	lateral face	solid
cone	net	sphere
cylinder	polygon	surface area
edge	prism	vertex
face	pyramid	volume
	slant height	

🎯 Identify the names and qualities of three-dimensional figures.
Sketch and draw solids in two dimensions and in three dimensions.
Describe two-dimensional figures that result from slicing three-dimensional figures.
Calculate the surface area of prisms.
Calculate the volume of prisms.
Calculate the surface area of pyramids.
Calculate the volume of pyramids.

Lesson 3.1 ~ Three-Dimensional Figures

Identify the number of faces, lateral faces, bases, edges and vertices in each solid.

1.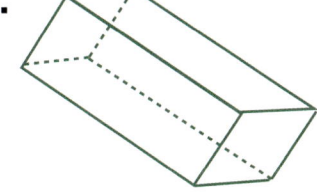

2.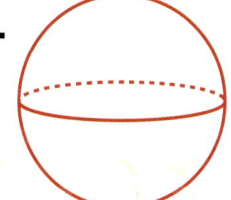

3.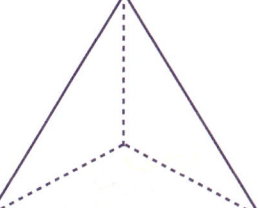

Name the solid that fits each description.

4. a can of hair spray

5. a basketball

6. a prism with 8 sides

7. a pyramid with 3 lateral faces

8. a solid with 8 vertices

9. a solid with two bases

Read each statement and determine if it is *always*, *sometimes* or *never* true. Explain your reasoning.

10. In a prism, the number of edges is always greater than the number of faces or vertices.

11. In a pyramid, the number of vertices is two more than the number of vertices on its base.

Block 3 ~ Review   119

Lesson 3.2 ~ Drawing Solids

12. What is a net?

Sketch a net of each solid.

13.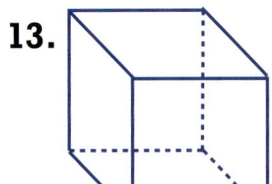

14.

15.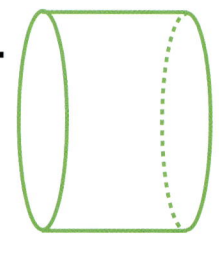

Sketch a diagram of each solid.

16. cone

17. trapezoidal prism

18. pentagonal pyramid

19. Draw two different nets for a cube.

Lesson 3.3 ~ Slicing Solids

20. A rectangular prism is 4 in by 6 in by 8 in. It is cut parallel to its base. The area of the rectangle created is 24 in^2. Which two dimensions form the base of the rectangular prism? Explain your reasoning.

21. Determine what shape is formed by each slice.

Solid	What shape is formed by a slice parallel to the base?	Is the parallel slice similar or congruent to the base?	What shape is formed by a perpendicular slice to the base?
Cylinder	a.	b.	c.
Cone	d.	e.	f.
Triangular Prism	g.	h.	i.
Octagonal Pyramid	j.	k.	l.

Name and sketch a diagram of the solid that matches each description.

22. Any slice to this solid will be a circle.

23. A parallel cut made to this solid is similar to its base and shaped like a triangle.

24. A slice made diagonally across this solid from the right side of the top base to the left side of the bottom base is an oval.

Lesson 3.4 ~ Surface Area of Prisms

Find the surface area of each prism.

25. The base of a cereal box is 14 square inches. The lateral area of the box is 162 square inches.

26. The lateral area of a hexagonal prism is 60 m^2. The base is 10 m^2.

Find the surface area of each prism.

27.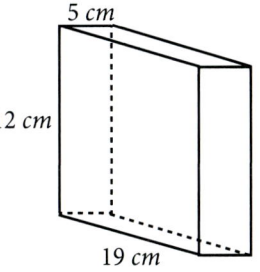

28.

29.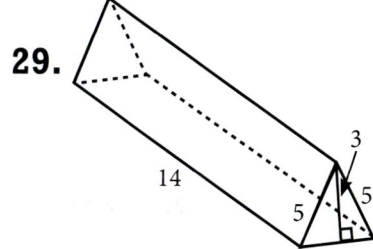

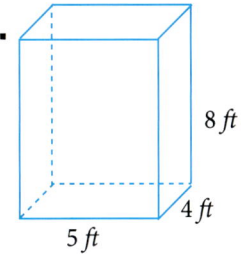

30. Kendra's toy chest needs new paint. The toy chest is 3 feet long, 1.5 feet wide and 2 feet tall. Find the amount of paint needed to cover the toy chest.

31. The surface area of a triangular prism is 35 in^2. The lateral area is 30 in^2. Find the area of one base. Use mathematics to justify your answer.

32. Jahzara took her state assessment test. In one problem, the dimensions of a box were 6 m by 10 m by 8 m. She used the formula $SA = 2(lw + wh + lh)$. Use this formula to calculate the surface area of the box.

Lesson 3.5 ~ Volume of Prisms

Find the volume of each prism.

33.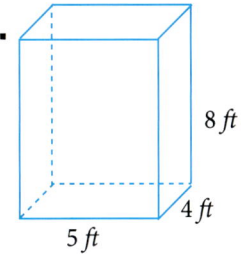

34.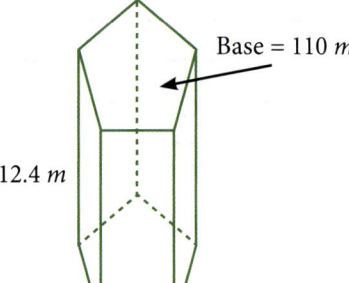

35.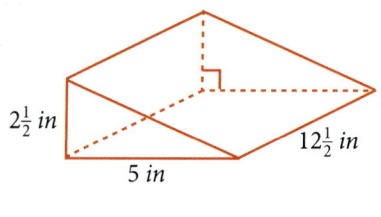

36. A square prism is 18 inches tall. The perimeter of the base is 40 inches. Find the volume of the prism.

37. Fernando went to the movies. He could get popcorn in a triangular container or a rectangular container. Each container was 10 inches tall. The sides of the rectangular container were 3 inches and 4 inches. The base of the triangular container had a side of 6 inches with a corresponding height of 5.2 inches. Which container held more popcorn? Show all work necessary to justify your answer.

38. The base of an octagonal hot tub is 77 ft^2. The hot tub is 3 feet deep.
   a. Find the volume of the hot tub.
   b. The manufacturer recommends only filling the hot tub $2\frac{1}{2}$ feet deep. What is the volume of water that should be put in the tub?

Lesson 3.6 ~ Surface Area of Regular Pyramids

39. How is the slant height of a pyramid different than the height of a pyramid?

40. Each side of a pentagonal pyramid is 7 centimeters. Find the lateral area of the pyramid if the slant height is 10 centimeters.

Find the surface area of each pyramid.

41.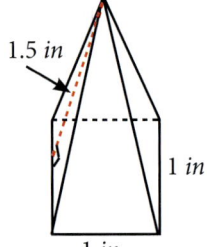
1.5 in, 1 in, 1 in

42.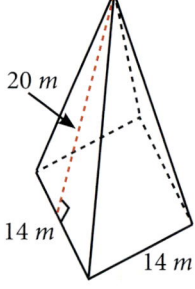
20 m, 14 m, 14 m

43. The Great Pyramid in Egypt was originally 754 feet on each side and the slant height was approximately 610 feet. Calculate the original lateral area of the Great Pyramid.

44. A square pyramid has a perimeter of 100 inches and a slant height of 10 inches. Find the surface area of the pyramid. Show all work necessary to justify your answer.

Lesson 3.7 ~ Volume of Pyramids

45. Describe the relationship between the volume of a pyramid and prism that have congruent bases and are the same height.

46. The volume of a pyramid is 22 cm^3. A prism has a congruent base and is the same height as the pyramid. What is the volume of the prism?

47. Sketch a diagram of a square pyramid. Identify the slant height (*l*) and the height (*h*) of the pyramid on your sketch.

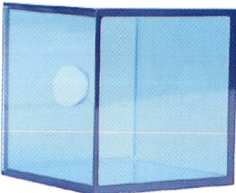

Find the volume of each pyramid.

48.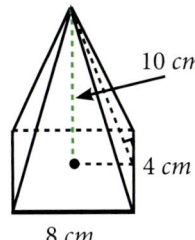

49. $h = 9$ in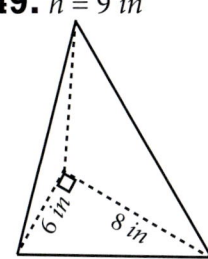

50.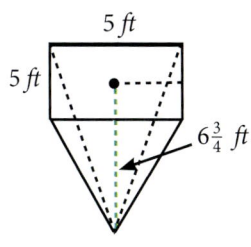

51. A candle shaped like a square pyramid has a base of 24 cm^2 and is 6 cm tall. What is the volume of the candle?

52. The volume of a square pyramid is 200 cubic centimeters. Each side of the square base has a length of 10 centimeters. What is the height of the pyramid? Show all work necessary to justify your answer.

Tic-Tac-Toe ~ Not Regular

The pyramids in **Lesson 3.6** have been regular pyramids.

1. Define regular pyramid.

2. Explain why a rectangular pyramid is not regular.

3. Draw a net of a rectangular pyramid. Cut out the net to make sure it forms a solid.

4. Measure needed lengths to the nearest tenth of a centimeter. Find the area of each figure in your net.

5. Find the surface area of your pyramid.

6. Explain why the formula for a regular pyramid (SA = $\frac{1}{2}Pl + B$) will not work for a rectangular pyramid.

7. Write a formula for calculating the surface area of a rectangular pyramid.

8. Clearly display all of your answers on a sheet of paper. Tape your net to your answer sheet.

CAREER FOCUS

RICK MASON

I am a mason. Masonry is one of the many trades that contribute to the completion of a construction project. Masonry projects include concrete block buildings and walls, brick-laying, stone working and fireplaces. I specialize in custom brick and stone work. My company contracts work with both general contractors and individual homeowners.

A masonry project requires the use of math from start to finish. I use basic math and Algebra skills so often that they become second nature. Calculating measurements, spacing and areas are a constant throughout my work day. Without accurate measurements, my finished work will not look good. Even before a project can begin, I use math skills to estimate the potential cost of doing a certain project.

Many masons have learned the trade by working with an experienced crew. Working with a crew gives you good on the job training and valuable experience. This is how I became a mason. Other masons attend community colleges or trade schools to learn how to do the work. Some unions even offer training and apprentice programs for people wanting to become masons.

Masons can expect to earn between $18,000 to $45,000 per year depending on expertise, experience and location. A mason may decide to work for a commercial contractor who builds large buildings, or they might work for a residential contractor and work on individual homes.

I enjoy my job because I am able to design and build a project that will be used and appreciated for generations. I welcome the challenge of working in a variety of settings from restoring historic buildings to stonework on vineyard estates. I also enjoy working in different locations like California, Montana or even Alaska. What I enjoy most about my job is seeing the results of my work at the end of each day.

ACKNOWLEDGEMENTS

**All Photos and Clipart ©2008 Jupiterimages Corporation and Clipart.com
with the exception of the cover photo and the following photos:**

Shapes & Angles Page 4
©iStockphoto.com/Alfonso Cacciola

Shapes & Angles Page 19
©iStockphoto.com/xyno

Shapes & Angles Page 47
©iStockphoto.com/Arno Staub

Shapes & Angles Page 59
©iStockphoto.com/Ivan Sizov

Shapes & Angles Page 64
©iStockphoto.com/xyno

Shapes & Angles Page 88
©iStockphoto.com/Branko Miokovic

Shapes & Angles Page 89
©iStockphoto.com/Massimo Merlini

Shapes & Angles Page 102
©iStockphoto.com/Tunisio Alves Filho

Shapes & Angles Page 106
©iStockphoto.com/Feng Yu

Shapes & Angles Page 110
©iStockphoto.com/design56

Shapes & Angles Page 117
©iStockphoto.com/Rafik El Raheb

Shapes & Angles Page 12
©iStockphoto.com/Aldo Murillo

Shapes & Angles Page 22
©iStockphoto.com/juanmonino

Shapes & Angles Page 58
©iStockphoto.com/Lora Clark

Shapes & Angles Page 60
©iStockphoto.com/Julián Rovagnati

Shapes & Angles Page 81
©iStockphoto.com/Katrina Brown

Shapes & Angles Page 89
©iStockphoto.com/Rob Sylvan

Shapes & Angles Page 99
©iStockphoto.com/Pgiam

Shapes & Angles Page 104
©iStockphoto.com/Claude Beaubien

Shapes & Angles Page 107
©iStockphoto.com/Kamil Macniak

Shapes & Angles Page 111
©iStockphoto.com/digitalskillet

Shapes & Angles Page 121
©iStockphoto.com/Stephanie Frey

Shapes & Angles Page 19
©iStockphoto.com/Aldo Murillo

Shapes & Angles Page 39
©iStockphoto.com/Kristian Septimius Krogh

Shapes & Angles Page 58
©iStockphoto.com/fstop123

Shapes & Angles Page 63
©iStockphoto.com/Jeanne Hatch

Shapes & Angles Page 83
©iStockphoto.com/Lisa F. Young

Shapes & Angles Page 89
©iStockphoto.com/keiran wills

Shapes & Angles Page 99
©iStockphoto.com/sunstock

Shapes & Angles Page 104
©iStockphoto.com/Chris Bernard

Shapes & Angles Page 109
©iStockphoto.com/Anthony Boulton

Shapes & Angles Page 114
©iStockphoto.com/Amanda Rohde

Layout and Design by Judy St. Lawrence

Cover Design by Schuyler St. Lawrence

Glossary Translation by Keyla Santiago and Heather Contreras

CORE FOCUS ON MATH
GLOSSARY ~ GLOSARIO

A

Absolute Value	The distance a number is from 0 on a number line.	**Valor Absoluto**	La distancia de un número desde el 0 en una recta numérica.
Acute Angle	An angle that measures more than 0° but less than 90°.	**Ángulo Agudo**	Un ángulo que mide mas 0° pero menos de 90°.
Adjacent Angles	Two angles that share a ray.	**Ángulos Adyacentes**	Dos ángulos que comparten un rayo.
Algebraic Expression	An expression that contains numbers, operations and variables.	**Expresiones Algebraicas**	Una expresión que contiene números, operaciones y variables.
Alternate Exterior Angles	Two angles that are on the outside of two lines and are on opposite sides of a transversal.	**Ángulos Exteriores Alternos**	Dos ángulos que están afuera de dos rectas y están a lados opuestos de una transversal.
Alternate Interior Angles	Two angles that are on the inside of two lines and are on opposites sides of a transversal.	**Ángulos Interiores Alternos**	Dos ángulos que están adentro de dos rectas y están a lados opuestos de una transversal.
Angle	A figure formed by two rays with a common endpoint.	**Ángulo**	Una figura formada por dos rayos con un punto final en común.

Area	The number of square units needed to cover a surface.	Área	El número de unidades cuadradas necesitadas para cubrir una superficie.
Ascending Order	Numbers arranged from least to greatest.	Progresión Ascendente	Los números ordenados de menor a mayor.
Associative Property	A property that states that numbers in addition or multiplication expressions can be grouped without affecting the value of the expression.	Propiedad Asociativa	Una propiedad que establece que los números en expresiones de suma o de multiplicación pueden ser agrupados sin afectar el valor de la expresión.
Axes	A horizontal and vertical number line on a coordinate plane. 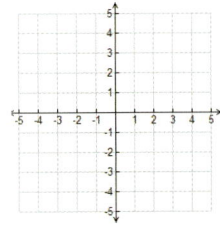	Ejes	Una recta numérica horizontal y vertical en un plano de coordenadas. 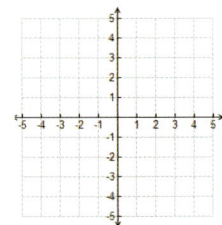
Axis of Symmetry	The line of symmetry on a parabola that goes through the vertex. 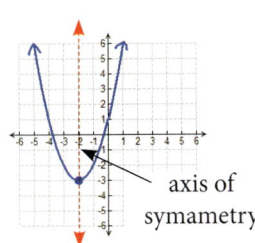	El Eje De Las Simetría	La linia de simetría de una parábola que pasa por el vértice. 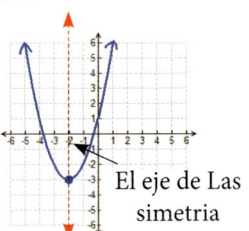

B

Bar Graph	A graph that uses bars to compare the quantities in a categorical data set. 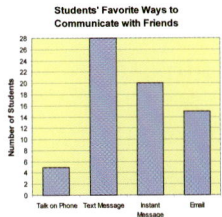	Gráfico de Barras	Una gráfica que utiliza barras para comparar las cantidades en un conjunto de datos categórico. 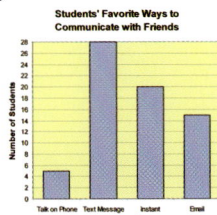
Base (of a power)	The base of the power is the repeated factor. In x^a, x is the base.	Base (de la potencia)	La base de la potenciación es el factor repatidio. En x^a, x es la base.

Glossary ~ Glosario

Base (of a solid)	See Prism, Cylinder, Pyramid and Cone.	Base (de un sólido)	Ver Prisma, Cilindro, Pirámide y Cono.
Base (of a triangle)	Any side of a triangle.	Base (de un triángulo)	Cualquier lado de un triángulo.
Bias	A problem when gathering data that affects the results of the data.	Sesgo	Un problema que ocurre cuando se recogen datos que afectan los resultados de los datos.
Biased Sample	A group from a population that does not accurately represent the entire population.	Muestra Sesgada	Un grupo de una población que no representa con exactitud la población entera.
Binomials	Expressions involving two terms (i.e. $x - 2$).	Binomiales	Expresiones que impliquen dos terminos. (es decir: $x - 2$).
Bivariate Data	Data that describes two variables and looks at the relationship between the two variables.	Datos de dos Variables	Los datos que describen dos variables y analiza la relación entre estas dos variables.
Box-and-Whisker Plot	A diagram used to display the five-number summary of a data set.	Diagrama de Líneas y Bloques	Un diagrama utilizado para mostrar el resumen de cinco números de un conjunto de datos.

C

Categorical Data	Data collected in the form of words.	Datos Categóricos	Datos recopilados en la forma de palabras.
Center of a Circle	The point inside a circle that is the same distance from all points on the circle.	Centro de un Círculo	Un ángulo dentro de un círculo que está a la misma distancia de todos los puntos en el círculo.

English	Definition	Spanish	Definición
Central Angle	An angle in a circle with its vertex at the center of a circle.	Ángulo Central	Un ángulo en un círculo con su vértice en el centro del círculo.
Chord	A line segment with endpoints on a circle.	Cuerda	Un segmento de la recta con puntos finales en el círculo.
Circle	The set of all points that are the same distance from a center point.	Círculo	El conjunto de todos los puntos que están a la misma distancia de un punto central.
Circumference	The distance around a circle.	Circunferencia	La distancia alrededor de un círculo.
Coefficient	The number multiplied by a variable in a term.	Coeficiente	El número multiplicado por una variable en un término.
Commutative Property	A property that states numbers can be added or multiplied in any order.	Propiedad Conmutativa	Una propiedad que establece que los números pueden ser sumados o multiplicados en cualquier orden.
Compatible Numbers	Numbers that are easy to mentally compute; used when estimating products and quotients.	Números Compatibles	Números que son fáciles de calcular mentalmente; utilizado cuando se estiman productos y cocientes.
Complementary Angles	Two angles whose sum is 90°.	Ángulos Complementarios	Dos ángulos cuya suma es de 90°.
Complements	Two probabilities whose sum is 1. Together they make up all the possible outcomes without repeating any outcomes.	Complementos	Dos probabilidades cuya suma es de 1. Juntos crean todos los posibles resultados sin repetir alguno.

Completing the Square	The creation of a perfect square trinomial by adding a constant to an expression in the form $x^2 + bx$.	Terminado el Cuadrado	La creación de un trinomio cuadrado perfecto por adición de una constante a una expresión en la forma $x^2 + bx$.
Complex Fraction	A fraction that contains a fractional expression in its numerator, denominator or both. $$\frac{\frac{3}{4}}{\frac{3}{8}}$$	Fracción Compleja	Una fracción que contiene una expresión fraccionaria en su numerador, el denominador o ambos. $$\frac{\frac{3}{4}}{\frac{3}{8}}$$
Composite Figure	A geometric figure made of two or more geometric shapes.	Figura Compuesta	Una figura geométrica formada por dos o más formas geométricas.
Composite Number	A whole number larger than 1 that has more than two factors.	Número Compuesto	Un número entero mayor que el 1 con más de dos factores.
Composite Solid	A solid made of two or more three-dimensional geometric figures.	Sólido Compuesto	Un sólido formado por dos o más figuras geométricas tridimensionales.
Composition of Transformations	A series of transformations on a point.	Composición de Transformaciones	Una serie de transformaciones en un punto.
Compound Probability	The probability of two or more events occurring.	Compuesto de Probabilidad	La probabilidad de dos o más eventos ocurriendo.
Conditional Frequency	The ratio of the observed frequency to the total number of frequencies in a given category from an experiment or survey.	Frecuencia Condicional	La relación de una frecuencia observada para el número total de frecuencias en una categoría dada del experimento o encuesta.
Cone	A solid formed by one circular base and a vertex.	Cono	Un sólido formado por una base circular y una vértice.
Congruent	Equal in measure.	Congruente	Igual en medida.

Congruent Figures	Two shapes that have the exact same shape and the exact same size. 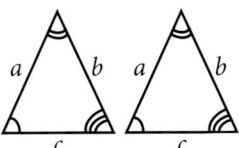	Figuras Congruentes	Dos figuras que tienen exactamente la misma forma y el mismo tamaño. 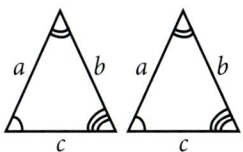
Constant	A term that has no variable.	Constante	Un término que no tiene variable.
Continuous	When a graph can be drawn from beginning to end without any breaks.	Continuo	Cuando una gráfica puede ser dibujada desde principio a fin sin ninguna interrupción.
Conversion	The process of renaming a measurement using different units.	Conversión	El proceso de renombrar una medida utilizando diferentes unidades.
Coordinate Plane	A plane created by two number lines intersecting at a 90° angle. 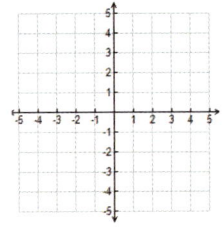	Plano de Coordenadas	Un plano creado por dos rectas numéricas que se intersecan a un ángulo de 90°. 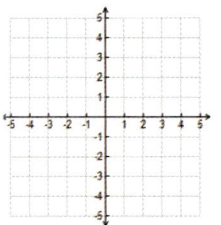
Correlation	The relationship between two variables in a scatter plot.	Correlación	La relación entre dos variables en un gráfico de dispersión.
Corresponding Angles	Two non-adjacent angles that are on the same side of a transversal with one angle inside the two lines and the other on the outside of the two lines. 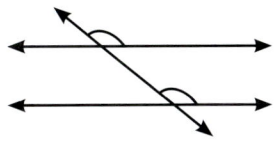	Ángulos Correspondientes	Dos ángulos no adyacentes que están en el mismo lado de una transversal con un ángulo adentro de las dos rectas y el otro afuera de las dos rectas. 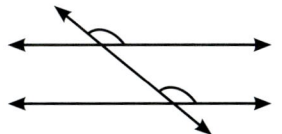
Corresponding Parts	The angles and sides in similar or congruent figures that match.	Partes Correspondientes	Los ángulos y lados en figuras similares o congruentes que concuerdan.

Glossary ~ Glosario

Cube Root	One of the three equal factors of a number. $3 \cdot 3 \cdot 3 = 27 \quad \sqrt[3]{27} = 3$	Raíz Cúbica	Uno de los tres factores iguales de un número. $3 \cdot 3 \cdot 3 = 27 \quad \sqrt[3]{27} = 3$
Cubed	A term raised to the power of 3.	Cubicado	Un término elevado a la potencia de 3.
Cylinder	A solid formed by two congruent and parallel circular bases.	Cilindro	Un sólido formado por dos bases circulares congruentes y paralelas.

D

Decimal	A number with a digit in the tenths place, hundredths place, etc.	Decimal	Un número con un dígito en las décimas, las centenas, etc.
Degrees	A unit used to measure angles.	Grados	Una unidad utilizada para medir ángulos.
Dependent Events	Two (or more) events such that the outcome of one event affects the outcome of the other event(s).	Eventos Dependiente	Dos (o más) eventos de tal manera que el resultado de un evento afecta el resultado del otro evento (s).
Dependent Variable	The variable in a relationship that depends on the value of the independent variable.	Variable Dependiente	La variable en una relación que depende del valor de la variable independiente.
Descending Order	Numbers arranged from greatest to least.	Progresión Descendente	Los números ordenados de mayor a menor.

132 Glossary ~ Glosario

Diameter	The distance across a circle through the center. 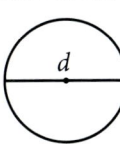	Diámetro	La distancia a través de un círculo por el centro. 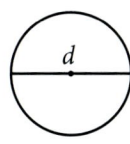
Dilation	A transformation which changes the size of the figure but not the shape. 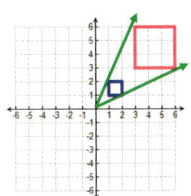	Dilatación	Una transformación que cambia el tamaño de la figura, pero no la forma. 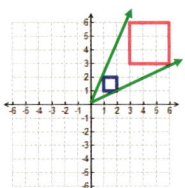
Direct Variation	A linear function that passes through the origin and has the equation $y = mx$.	Variación Directa	Una función lineal que pasa a través del origen y tiene la ecuación $y = mx$.
Discount	The decrease in the price of an item.	Descuento	La disminución de precio en un artículo.
Discrete	When a graph can be represented by a unique set of points rather than a continuous line.	Discreto	Cuando una gráfica puede ser representada por un conjunto de puntos único en vez de una recta continua.
Discriminant	In the quadratic formula, the expression under the radical sign. The discriminant provides information about the number of real roots or solutions of a quadratic equation. $$\frac{-b \pm \sqrt{b^2 - 4ac}}{2a}$$	Discriminante	En la fórmula cuadrática, la expresión bajo el signo radical. El discriminante proporciona información sobre el número o las verdaderas raíces o soluciones de una ecuación cuadrática. $$\frac{-b \pm \sqrt{b^2 - 4ac}}{2a}$$
Distance Formula	A formula used to find the distance between two points on the coordinate plane. $$d = \sqrt{(x_2 - x_1)^2 + (y_2 - y_1)^2}$$	Fórmula de Distancia	Una fórmula utilizada para encontrar la distancia entre dos puntos en un plano de coordenadas. $$d = \sqrt{(x_2 - x_1)^2 + (y_2 - y_1)^2}$$

Distributive Property	A property that can be used to rewrite an expression without parentheses. $a(b+c) = ab + ac$	Propiedad Distributiva	Una propiedad que puede ser utilizada para reescribir una expresión sin paréntesis: $a(b+c) = ab + ac$
Dividend	The number being divided. $\mathbf{100} \div 4 = 25$	Dividendo	El número que es dividido. $\mathbf{100} \div 4 = 25$
Divisor	The number used to divide. $100 \div \mathbf{4} = 25$	Divisor	El número utilizado para dividir. $100 \div \mathbf{4} = 25$
Domain	The set of input values of a function.	Dominio	El conjunto de valores entrados de la función.
Dot Plot	A data display that consists of a number line with dots equally spaced above data values. 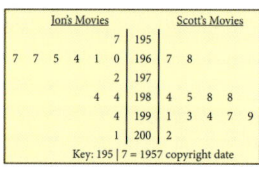	Punto de Gráfico	Una visualización de datos que consiste de una línea numérica con puntos igualmente espaciados sobre valores de datos.
Double Stem-and-Leaf Plot	A stem-and-leaf plot where one set of data is placed on the right side of the stem and another is placed on the left of the stem.	Doble Gráfica de Tallo y Hoja	Una gráfica de tallo y hoja donde un conjunto de datos es colocado al lado derecho del tallo y el otro es colocado a la izquierda del tallo.

E

Edge	The segment where two faces of a solid meet. 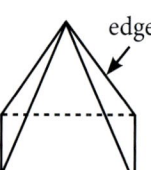	Arista (Borde)	El segmento donde dos caras de un sólido se encuentran. 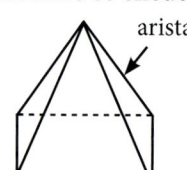

English	Definition	Spanish	Definición
Elimination Method	A method for solving a system of linear equations.	Método de Eliminación	Un método para resolver un sistema de ecuaciones lineales.
Enlargement	A dilation that creates an image larger than its pre-image.	Agrandamiento	Una dilatación que crea una imagen más grande que su pre-imagen.
Equally Likely	Two or more possible outcomes of a given situation that have the same probability.	Igualmente Probables	Dos o más posibles resultados de una situación dada que tienen la misma probabilidad.
Equation	A mathematical sentence that contains an equals sign between 2 expressions.	Ecuación	Una oración matemática que contiene un símbolo de igualdad entre dos expresiones.
Equiangular	A polygon in which all angles are congruent.	Equiángulo	Un polígono en el cual todos los ángulos son congruentes.
Equilateral	A polygon in which all sides are congruent.	Equilátero	Un polígono en el cual todos los lados son congruentes.
Equivalent Decimals	Two or more decimals that represent the same number.	Decimales Equivalentes	Dos o más decimales que representan el mismo número.
Equivalent Expressions	Two or more expressions that represent the same algebraic expression.	Expresiones Equivalentes	Dos o más expresiones que representan la misma expresión algebraica.
Equivalent Fractions	Two or more fractions that represent the same number.	Fracciones Equivalentes	Dos o más fracciones que representan el mismo número.
Evaluate	To find the value of an expression.	Evaluar	Encontrar el valor de una expresión.
Even Distribution	A set of data values that is evenly spread across the range of the data.	Distribución Igualada	Un conjunto de valores de datos que es esparcido de modo uniforme a través del rango de los datos.

English	Definition	Spanish	Definición
Event	A desired outcome or group of outcomes.	Evento	Un resultado o grupo de resultados deseados.
Experimental Probability	The ratio of the number of times an event occurs to the total number of trials.	Probabilidad Experimental	La razón de la cantidad de veces que un suceso ocurre a la cantidad total de intentos.
Exponent	In x^a, a is the exponent. The exponent shows the number of times the factor (x) is repeated.	Exponente	En x^a, a es el exponente. El exponente indica el número de veces que se repite el factor (x).
Exponential Function	A function that can be described by an equation in the form $f(x) = bm^x$.	Función Exponencial	Una función que puede ser descrito por una ecuación en la forma $f(x) = bm^x$.

F

English	Definition	Spanish	Definición
Face	A polygon that is a side or base of a solid.	Cara	Un polígono que es una base de lado de un sólido.
Factors	Whole numbers that can be multiplied together to find a product.	Factores	Números enteros que pueden ser multiplicados entre si para encontrar un producto.
First Quartile (Q1)	The median of the lower half of a data set.	Primer Cuartil (Q1)	Mediana de la parte inferior de un conjunto de datos.
Five-Number Summary	Describes the spread of a data set using the minimum, 1st quartile, median, 3rd quartile and maximum.	Sumario de Cinco Números	Describe la extensión de un conjunto de datos utilizando el mínimo, el primer cuartil, la mediana el tercer cuartil y el máximo.
Formula	An algebraic equation that shows the relationship among specific quantities.	Fórmula	Una ecuación algebraica que enseña la relación entre cantidades específicas.
Fraction	A number that represents a part of a whole number, written as $\frac{numerator}{denominator}$.	Fracción	Un número que representa una parte de un número entero, escrito como $\frac{numerador}{denominador}$.

English	Definition	Spanish	Definición
Frequency	The number of times an item occurs in a data set.	Frecuencia	La cantidad de veces que un artículo ocurre en un conjunto de datos.
Frequency Table	A table which shows how many times a value occurs in a given interval.	Tabla de Frecuencia	Una tabla que enseña cuantas veces un valor ocurre en un intervalo dado.
Function	A relationship between two variables that has one output value for each input value.	Función	Una relación entre dos variables que tiene un valor de salida para cada valor de entrada.

G

English	Definition	Spanish	Definición
General Form	A quadratic function is in general form when written $f(x) = ax^2 + bx + c$ where $a \neq 0$.	Forma General	Una función cuadrática es en forma general cuándo escrito $f(x) = ax^2 + bx + c$ donde $a \neq 0$.
Geometric Probability	Ratios of lengths or areas used to find the likelihood of an event.	Probabilidad Geométrica	Razones de longitudes o áreas utilizadas para encontrar la probabilidad de un suceso.
Geometric Sequence	A list of numbers that begins with a starting value. Each term in the sequence is generated by multiplying the previous term in the sequence by a constant multiplier.	Secuenciación Geométrica	Una lista de números que comienza con un valor inicial. Cada término de la secuencia se genera al multiplicar el término anterior de la secuencia por un multiplicar constante.
Greatest Common Factor (GCF)	The greatest factor that is common to two or more numbers.	Máximo Común Divisor (MCD)	El máximo divisor que le es común a dos o más números.
Grouping Symbols	Symbols such as parentheses or fraction bars that group parts of an expression.	Símbolos de Agrupación	Símbolos como el paréntesis o barras de fracción que agrupan las partes de una expresión.

H

Height of a Triangle	A perpendicular line drawn from the side whose length is the base to the opposite vertex. 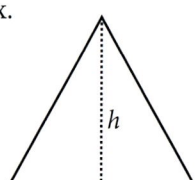	Altura de un Triángulo	Una recta perpendicular dibujada desde el lado cuya longitud es la base al vértice opuesto. 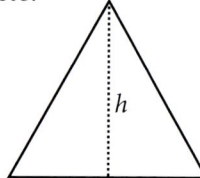
Histogram	A bar graph that displays the frequency of numerical data in equal-sized intervals. 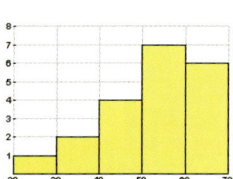	Histograma	Un gráfico de barras que muestra la frecuencia de datos numéricos en intervalos de tamaños iguales. 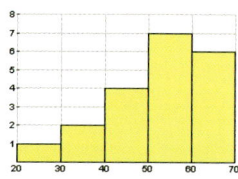
Hypotenuse	The side opposite the right angle in a right triangle. 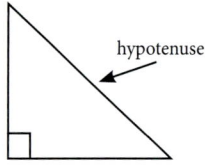	Hipotenusa	El lado opuesto el ángulo recto en un triángulo rectángulo. 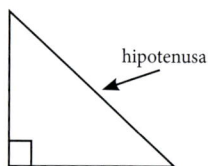

I-J-K

Image	A point or figure which is the result of a transformation or series of transformations.	Imagen	Un punto o figura que es el resultado de una transformación o una serie de transformaciones.
Improper Fraction	A fraction whose numerator is greater than or equal to its denominator.	Fracción Impropia	Una fracción cuyo numerador es mayor o igual a su denominador.
Independent Events	Two (or more) events such that the outcome of one event does not affect the outcome of the other event(s).	Eventos Independientes	Dos (o más) eventos de tal manera que el resultado de un evento no afecta el resultado del otro evento (s).

English	Definition	Español	Definición
Independent Variable	The variable representing the input values. ←independent	Variable Independiente	La variable que representa los valores entratos. ←independiente
Inequality	A mathematical sentence using <, >, ≤ or ≥ to compare two quantities.	Desigualdad	Un enunciado matemático usando <, >, ≤ ó ≥ para comparar dos cantidades.
Inference	A logical conclusion based on known information.	Inferencia	Una conclusión lógica basada en la información conocida.
Input-Output Table	A table used to describe a function by listing input values with their output values.	Tabla de Entrada y Salida	Una tabla utilizada para describir una función al enumerar valores de entrada con sus valores de salidas.
Integers	The set of all whole numbers, their opposites, and 0.	Enteros	El conjunto de todos los números enteros, sus opuestos y 0.
Interquartile Range (IQR)	The difference between the 3rd quartile and the 1st quartile in a set of data.	Rango Intercuartil (IQR)	La diferencia entre el tercer cuartil y el primer cuartil en un conjunto de datos.
Inverse Operations	Operations that undo each other.	Operaciones Inversas	Operaciones que se cancelan la una a la otra.
IQR Method	A method for determining outliers using interquartile ranges.	Método IQR	Un método para determinar los datos aberrantes.
Irrational Numbers	A number that cannot be expressed as a fraction of two integers.	Números Irracionales	Un número que no puede ser expresado como una fracción de dos enteros.

Glossary ~ Glosario

Isosceles Trapezoid	A trapezoid that has congruent legs. 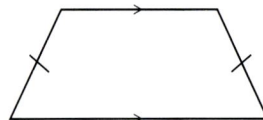	Trapezoide Isósceles	Un trapezoide con catetos congruentes. 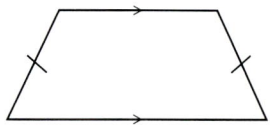
Isosceles Triangle	A triangle that has two or more congruent sides. 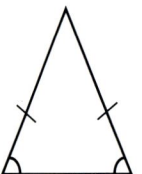	Triángulo Isósceles	Un triángulo que tiene dos o más lados congruentes. 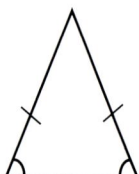

L

Lateral Face	A side of a solid that is not a base.	Cara Lateral	Un lado de un sólido que no sea una base.
Least Common Denominator (LCD)	The least common multiple of two or more denominators.	Mínimo Común Denominador (MCD)	El mínimo común múltiplo de dos o más denominadores.
Least Common Multiple (LCM)	The smallest nonzero multiple that is common to two or more numbers.	Mínimo Común Múltiplo (MCM)	El múltiplo más pequeño que no sea cero que le es común a dos o más números.
Leg	The two sides of a right triangle that form a right angle.	Cateto	Los dos lados de un triángulo rectángulo que forman un ángulo recto.
Like Terms	Terms that have the same variable(s).	Términos Semejantes	Términos que tienen el mismo variable(s).

Line of Best Fit	A line which best represents the pattern of a two-variable data set.	Recta de Mejor Ajuste	Una recta que mejor representa el patrón de un conjunto de datos de dos variables.
Linear Equation	An equation whose graph is a line.	Ecuación Lineal	Una ecuación cuya gráfica es una recta.
Linear Function	A function whose graph is a line.	Función Lineal	Una función cuya gráfica es una recta.
Linear Pair	Two adjacent angles whose non-common sides are opposite rays.	Par Lineal	Dos ángulos adyacentes cuyos lados no comunes son rayos opuestos.

M

Mark-up	The increase in the price of an item.	Margen de Beneficio	El aumento de precio en un artículo.
Maximum	The highest point on a curve.	Máximo	El punto más alto en la curva.
Mean	The sum of all values in a data set divided by the number of values.	Media	La suma de todos los valores en un conjunto de datos dividido entre la cantidad de valores.
Mean Absolute Deviation	A statistic that shows the average distance from the mean for all numbers in a data set.	Desviación Media Absoluta	Una estadística que muestra la distancia promedio entre la media de todos los números en una serie de datos.

Glossary ~ Glosario

Measures of Center	Numbers that are used to represent a data set with a single value; the mean, median, and mode are the measures of center.	Medidas de Centro	Números que son utilizados para representar un conjunto de datos con un solo valor; la media, la mediana, y la moda son las medidas de centro.
Measures of Variability	Statistics that help determine the spread of numbers in a data set.	Medidas de Variabilidad	Las estadísticas que ayudan a determinar la extensión de los números en una serie de datos.
Median	The middle number or the average of the two middle numbers in an ordered data set.	Mediana	El número medio o el promedio de los dos números medios en un conjunto de datos ordenados.
Minimum	The lowest point on a curve.	Mínimo	El punto más bajo en la curva.
Mixed Number	The sum of a whole number and a fraction less than 1.	Números Mixtos	La suma de un número entero y una fracción menor que 1.
Mode	The number(s) or item(s) that occur most often in a data set.	Moda	El número(s) o artículo(s) que ocurre con más frecuencia en un conjunto de datos.
Motion Rate	A rate that compares distance to time.	Índice de Movimiento	Un índice que compara distancia por tiempo.
Multiple	The product of a number and nonzero whole number.	Múltiplo	El producto de un número y un número entero que no sea cero.

N

Negative Number	A number less than 0.	Número Negativo	Un número menor que 0.

Net	A two-dimensional pattern that folds to form a solid.	Red	Un patrón bidimensional que se dobla para formar un sólido.
Non-Linear Function	A function whose graph does not form a line.	Ecuación No Lineal	Una ecuación cuya gráfica no forma una recta.
Normal Distribution	A set of data values where the majority of the values are located in the middle of the data set and can be displayed by a bell-shaped curve.	Distribución Normal	Un conjunto de valores de datos donde la mayoría de los valores están localizados en el medio del conjunto de datos y pueden ser mostrados por una curva de forma de campana.
Numerical Data	Data collected in the form of numbers.	Datos Numéricos	Datos recopilados en la forma de números.
Numerical Expressions	An expression consisting of numbers and operations that represents a specific value.	Expresiones Numéricas	Una expresión que consta de números y operaciones que representa un valor específico.

O

Obtuse Angle	An angle that measures more than 90° but less than 180°.	Ángulo Obtuso	Un ángulo que mide más de 90° pero menos de 180°.
Opposites	Numbers that are the same distance from 0 on a number line but are on opposite sides of 0.	Opuestos	Números a la misma distancia del 0 en un recta numérica pero en lados opuestos del 0.
Order of Operations	The rules to follow when evaluating an expression with more than one operation.	Orden de Operaciones	Las reglas a seguir cuando se evalúa una expresión con más de una operación.
Ordered Pair	A pair of numbers used to locate a point on a coordinate plane (x, y).	Par Ordenado	Un par de números utilizados para localizar un punto en un plano de coordenadas (x, y).

Origin	The point where the *x*- and *y*-axis intersect on a coordinate plane (0, 0). 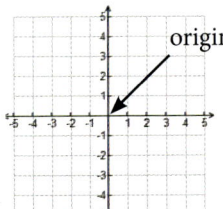	Origen	El punto donde el eje de la *x*- *y* el de la *y*- se cruzan en un plano de coordinadas (0,0). 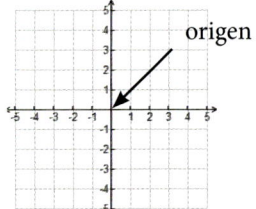
Outcome	One possible result from an experiment or a probability sample space.	Resultado	Un resultado posible de un experimento o un espacio de probabilidad de la muestra.
Outlier	An extreme value that varies greatly from the other values in a data set.	Dato Aberrante	Un valor extremo que varía mucho de los otros valores en un conjunto de datos.

P

Parabola	The graph of a quadratic function. 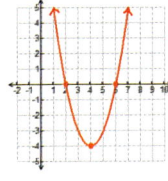	Parábola	La gráfica de una función cuadratica. 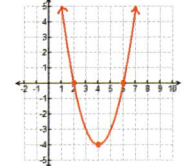
Parallel	Lines in the same plane that never intersect.	Paralela	Rectas en el mismo plano que nunca se intersecan.
Parallel Box-and-Whisker Plot	One box-and-whisker plot placed above another; often used to compare data sets. 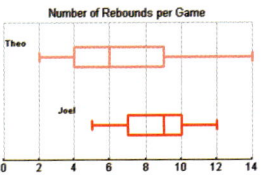	Diagrama Paralelo de Líneas y Bloques	Un diagrama de líneas y bloques ubicado sobre otro para comparar conjuntos de datos. 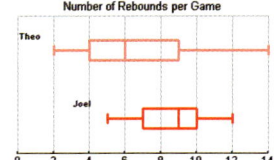

Parallelogram	A quadrilateral with both pairs of opposite sides parallel.	Paralelogramo	Un cuadrilateral con ambos pares de lados opuestos paralelos.
Parent Function	The simplest form of a particular type of function.	Función Principal	La forma más simple de un tipo particular de la función.
Parent Graph	The most basic graph of a function.	Gráfico Matriz	La gráfica más básica de una función.
Percent	A ratio that compares a number to 100.	Por Ciento	Una razón que compara un número con 100.
Percent of Change	The percent a quantity increases or decreases compared to the original amount.	Por Ciento de Cambio	El por ciento que una cantidad aumenta o disminuye comparado a la cantidad original.
Percent of Decrease	The percent of change when the new amount is less than the original amount.	Por Ciento de Disminución	El por ciento de cambio cuando la nueva cantidad es menos que la cantidad original.
Percent of Increase	The percent of change when the new amount is more than the original amount.	Por Ciento de Incremento	El por ciento de cambio cuando la nueva cantidad es más que la cantidad original.
Perfect Cube	A number whose cube root is an integer.	Cubo Perfecto	Un número cuyo raíz cúbica es un número entero.
Perfect Square	A number whose square root is an integer.	Cuadrado Perfecto	Un número cuyo raíz cuadrado es un número entero.
Perfect Square Trinomial	A trinomial that is the square of a binomial.	Trinomio Cuadrado Perfecto	Un trinomio que es el cuadrado de un binomio.
Perimeter	The distance around a figure.	Perímetro	La distancia alrededor de una figura.

Perpendicular	Two lines or segments that form a right angle.	Perpendicular	Dos rectas o segmentos que forman un ángulo recto.
Pi (π)	The ratio of the circumference of a circle to its diameter.	Pi (π)	La razón de la circunferencia de un círculo a su diámetro.
Pictograph	A graph that uses pictures to compare the amounts represented in a categorical data set.	Gráfica Pictórica	Una gráfica que utiliza dibujos para comparar las cantidades representadas en un conjunto de datos categóricos.
Pie Chart	A circle graph that shows information as sectors of a circle.	Gráfico Circular	Enseña la información como sectores de un círculo.
Polygon	A closed figure formed by three or more line segments.	Polígono	Una figura cerrada formada por tres o más segmentos de rectas.
Population	The entire group of people or objects one wants to gather information about.	Población	Todo el grupo de personas o los objetos a los que se quiere obtener información sobre.
Positive Number	A number greater than 0.	Número Positivo	Un número mayor que 0.
Power	An expression such as x^a which consists of two parts, the base (x) and the exponent (a).	Potencia	Una expresión como x^a que consiste de dos partes, la base (x) y el exponente (a).
Pre-image	The original figure prior to a transformation.	Pre-imagen	La figura original antes de una transformación.

Prime Factorization	When any composite number is written as the product of all its prime factors.	**Factorización Prima**	Cuando cualquier número compuesto es escrito como el producto de todos los factores primos.
Prime Number	A whole number larger than 1 that has only two possible factors, 1 and itself.	**Número Primo**	Un número entero mayor que 1 que tiene solo dos factores posibles, 1 y el mismo.
Prism	A solid formed by polygons with two congruent, parallel bases.	**Prisma**	Un sólido formado por polígonos con dos bases congruentes y paralelas.
Probability	The measure of how likely it is an event will occur.	**Probabilidad**	La medida de cuán probable un suceso puede ocurrir.
Product	The answer to a multiplication problem.	**Producto**	La respuesta a un problema de multiplicación.
Proper Fraction	A fraction with a numerator that is less than the denominator.	**Fracción Propia**	Una fracción con un numerador que es menos que el denominador.
Proportion	An equation stating two ratios are equivalent.	**Proporción**	Una ecuación que establece que dos razones son equivalentes.
Protractor	A tool used to measure angles.	**Transportador**	Una herramienta para medir ángulos.
Pyramid	A solid with a polygonal base and triangular sides that meet at a vertex.	**Pirámide**	Un sólido con una base poligonal y lados triangulares que se encuentran en un vértice.

Pythagorean Triple	A set of three positive integers (*a*, *b*, *c*) such that $a^2 + b^2 = c^2$.	Triple de Pitágoras	Un conjunto de tres enteros positivos (*a*, *b*, *c*) tal que $a^2 + b^2 = c^2$.

Q

Q-Points	Points that are created by the intersection of the quartiles for the *x*- and *y*-values of a two-variable data set.	Puntos Q	Puntos que son creados por la intersección de los cuartiles para los valores de la *x*- y la *y*- de un conjunto de datos de dos variables.
Quadrants	Four regions formed by the *x* and *y* axes on a coordinate plane. 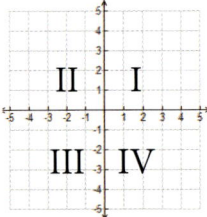	Cuadrantes	Cuatro regiones formadas por el eje-*x* y el eje-*y* en un plano de coordinadas. 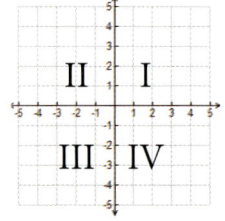
Quadratic Formula	A method which can be used to solve quadratic equations in the form $0 = ax^2 + bx + c$, where $a \neq 0$. $$x = \frac{-b \pm \sqrt{b^2 - 4ac}}{2a}$$	Fórmula Cuadrática	Un métado que puede usarse para resolver ecuaciones cuadraticas en la forma $0 = ax^2 + bx + c$, donde $a \neq 0$. $$x = \frac{-b \pm \sqrt{b^2 - 4ac}}{2a}$$
Quadratic Function	Any function in the family with the parent function of $f(x) = x^2$.	Función Cuadrática	Cualquier otra función en la familia con la función principal de $f(x) = x^2$.
Quadrilateral	A polygon with four sides.	Cuadrilateral	Un polígono con cuatro lados.
Quotient	The answer to a division problem.	Cociente	La solución a un problema de división.

R

Radius	The distance from the center of a circle to any point on the circle. 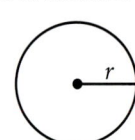	Radio	La distancia desde el centro de un círculo a cualquier punto en el círculo.

Random Sample	A group from a population created when each member of the population is equally likely to be chosen.	Muestra Aleatoria	Un grupo de una población creada cuando cada miembro de la población tiene la misma probabilidad de ser elegido.
Range (of a data set)	The difference between the maximum and minimum values in a data set.	rango	La diferencia entre los valores máximo y mínimo en un conjunto de datos.
Range (of a function)	The set of output values of a function.	Rango (de una función)	El conjunto de valores salidos de la función.
Rate	A ratio of two numbers that have different units.	Índice	Una proporción de dos números con diferentes unidades.
Rate Conversion	A process of changing at least one unit of measurement in a rate to a different unit of measurement.	Conversión de Índice	Un proceso de cambiar por lo menos una unidad de medición en un índice a una diferente unidad de medición.
Rate of Change	The change in y-values over the change in x-values on a linear graph.	Índice de Cambio	El cambio en los valores de y sobre el cambio en los valores de x en una gráfica lineal.
Ratio	A comparison of two numbers using division. $a:b \quad \frac{a}{b} \quad a \text{ to } b$	Razón	Una comparación de dos números utilizando división. $a:b \quad \frac{a}{b} \quad a \text{ a } b$
Rational Number	A number that can be expressed as a fraction of two integers.	Número Racional	Un número que puede ser expresado como una fracción de dos enteros.
Ray	A part of a line that has one endpoint and extends forever in one direction.	Rayo	Una parte de una recta que tiene un punto final y se extiende eternamente en una dirección.
Real Numbers	The set of numbers that includes all rational and irrational numbers.	Números Racionales	El conjunto de números que incluye todos los números racionales e irracionales.

Reciprocals	Two numbers whose product is 1.	Recíprocos	Dos números cuyo producto es 1.
Rectangle	A quadrilateral with four right angles.	Rectángulo	Un cuadrilátero con cuatro ángulos rectos.
Recursive Routine	A routine described by stating the start value and the operation performed to get the following terms.	Rutina Recursiva	Una rutina descrita al exponer el valor del comienzo y la operación realizada para conseguir los términos siguientes.
Recursive Sequence	An ordered list of numbers created by a first term and a repeated operation.	Secuencia Recursiva	Una lista de números ordenados creada por un primer término y una operación repetida.
Reduction	A dilation that creates an image smaller than its pre-image.	Reducción	Una dilatación que crea una imagen más pequeña que su pre-imagen.
Reflection	A transformation in which a mirror image is produced by flipping a figure over a line.	Reflexión	Una transformación en el que se produce una imagen reflejada volteando una figura sobre una línea.
Relative Frequency	The ratio of the observed frequency to the total number of frequencies in an experiment or survey.	Frecuencia Relativa	La proporción de la frecuencia observada para el número total de frecuencias en un experimento o estudio.
Remainder	A number that is left over when a division problem is completed.	Remanente	Un número que queda cuando un problema de división se ha completado.
Repeating Decimal	A decimal that has one or more digits that repeat forever.	Decimal Repetitivo	Un decimal que tiene uno o más dígitos que se repiten eternamente.

Representative Sample	A group from a population that accurately represents the entire population.	**Muestra Representativa**	Un grupo de una población que representa con precisión toda la población.
Rhombus	A quadrilateral with four sides equal in measure. 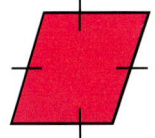	**Rombo**	Un cuadrilátero con cuatro lados iguales en la medida. 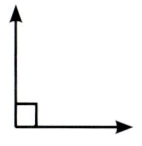
Right Angle	An angle that measures 90°. 	**Ángulo Recto**	Un ángulo que mide 90°.
Roots	The x-intercepts of a quadratic function. 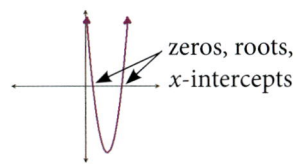	**Raíces**	Las intersecciones-x de una función cuadratica. 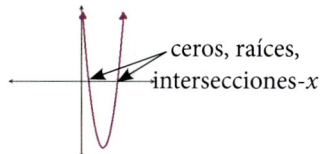
Rotation	A transformation which turns a point or figure about a fixed point, often the origin. 	**Rotación**	Una transformación que convierte un punto a una figura sobre un punto fijo. 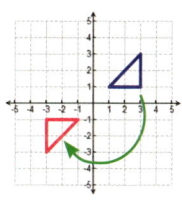

S

Sales Tax	An amount added to the cost of an item. The amount added is a percent of the original amount as determined by a state, county or city.	**Impuesto sobre las Ventas**	Una cantidad añadida al costo de un artículo. La cantidad añadida es un por ciento de la cantidad original determinado por el estado, condado o ciudad.

Same-Side Interior Angles	Two angles that are on the inside of two lines and are on the same side of a transversal. 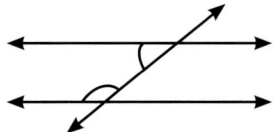	Ángulos Interiores del Mismo Lado	Dos ángulos que están en el interior de dos rectas y están en el mismo lado de una transversal. 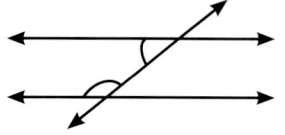
Sample	A group from a population that is used to make conclusions about the entire population.	Muestra	Un grupo de una población que se utiliza para sacar conclusiones sobre toda la población.
Sample Space	The set of all possible outcomes.	Muestra de Espacio	El conjunto de todos los resultados posibles.
Scale	The ratio of a length on a map or model to the actual object.	Escala	La razón de una longitud en un mapa o modelo al objeto verdadero.
Scale Factor	The ratio of corresponding sides in two similar figures.	Factor de Escala	La razón de los lados correspondientes en dos figuras similares.
Scalene Triangle	A triangle that has no congruent sides. 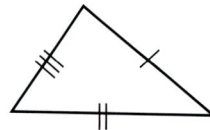	Triángulo Escaleno	Un triángulo sin lados congruentes. 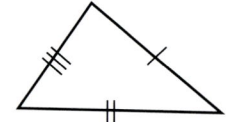
Scatter Plot	A set of ordered pairs graphed on a coordinate plane. 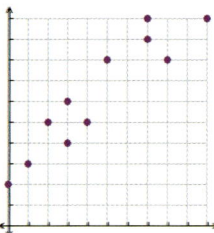	Diagrama de Dispersión	Un conjunto de pares ordenados graficados en un plano de coordenadas. 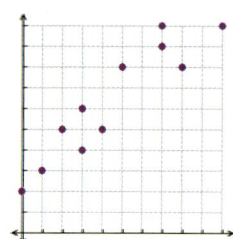
Scientific Notation	Scientific notation is an exponential expression using a power of 10 where $1 \leq N < 10$ and P is an integer. $N \times 10^P$	La Notación Científica	Notación científica es una expresión exponencial con una potencia de 10, donde $1 \leq N < 10$ y P es un número entero. $N \times 10^P$

Sector	A portion of a circle enclosed by two radii.	Sector	Una porción de un circulo encerado por dos radios.
Sequence	An ordered list of numbers.	Sucesión	Una lista de números ordenados.
Similar Figures	Two figures that have the exact same shape, but not necessarily the exact same size.	Figuras Similares	Dos figuras que tienen exactamente la misma forma, pero no necesariamente el mismo tamaño exacto.
Similar Solids	Solids that have the same shape and all corresponding dimensions are proportional.	Sólidos Similares	Sólidos con la misma forma y todas sus dimensiones correspondientes son proporcionales.
Simplest Form	A fraction whose numerator and denominator's only common factor is 1.	Expresión Simple	Una fracción cuyo único factor común del numerador y del denominador es 1.
Simplify an Expression	To rewrite an expression without parentheses and combine all like terms.	Simplificar una Expresión	Reescribir una expresión sin paréntesis y combinar todos los términos iguales.
Simulation	An experiment used to model a situation.	Simulación	Un experimento utilizado para modelar una situación.
Single-Variable Data	A data set with only one type of data.	Datos de una Variable	Un conjunto de datos con tan solo un tipo de datos.
Sketch	To make a figure free hand without the use of measurement tools.	Esbozo	Hacer una figura a mano libre sin utilizar herramientas de medidas.
Skewed Left	A plot or graph with a longer tail on the left-hand side.	Torcido a la Izquierda	Un gráfico con una cola al lado izquierdo.

Glossary ~ Glosario **153**

Skewed Right	A plot or graph with a longer tail on the right-hand side.	**Torcido a la Derecha**	Un gráfico con una cola al lado derecho.
Slant Height	The height of a lateral face of a pyramid or cone.	**Altura Sesgada**	La altura de un cara lateral de una pirámide o cono.
Slope	The ratio of the vertical change to the horizontal change in a linear graph.	**Pendiente**	La razón del cambio vertical al cambio horizontal en una gráfica lineal.
Slope Triangle	A right triangle formed where one leg represents the vertical rise and the other leg is the horizontal run in a linear graph.	**Triángulo de Pendiente**	Un triángulo rectángulo formado donde una cateto representa el ascenso y la otra es una carrera horizontal en una gráfica lineal.
Slope-Intercept Form	A linear equation written in the form $y = mx + b$.	**Forma de las Intersecciones con la Pendiente**	Una ecuación lineal escrita en la forma $y = mx + b$.
Solid	A three-dimensional figure that encloses a part of space.	**Sólido**	Una figura tridimensional que encierra una parte del espacio.
Solution	Any value or values that makes an equation true.	**Solución**	Cualquier valor o valores que hacen una ecuación verdadera.
Solution of a System of Linear Equations	The ordered pair that satisfies both linear equations in the system.	**Solución de un Sistema de Ecuaciones Lineales**	El par ordenado que satisface ambas ecuaciones lineales en el sistema.

English	Definition	Español	Definición
Sphere	A solid formed by a set of points in space that are the same distance from a center point. 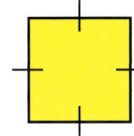	Esfera	Un sólido formado por un conjunto de puntos en el espacio que están a la misma distancia de un punto central. 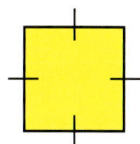
Square	A quadrilateral with four right angles and four congruent sides.	Cuadrado	Un cuadrilátero con cuatro ángulos rectos y cuatro lados congruente.
Square Root	One of the two equal factors of a number. $3 \cdot 3 = 9 \quad 3 = \sqrt{9}$	Raíz Cuadrada	Uno de los dos factores iguales de un número. $3 \cdot 3 = 9 \quad 3 = \sqrt{9}$
Squared	A term raised to the power of 2.	Cuadrado	Un término elevado a la potencia de 2.
Start Value	The output value that is paired with an input value of 0 in an input-output table.	Valor de Comienzo	El valor de salida que es aparejado con un valor de entrada de 0 en una tabla de entradas y salidas.
Statistics	The process of collecting, displaying and analyzing a set of data.	Estadísticas	El proceso de recopilar, exponer y analizar un conjunto de datos.
Stem-and-Leaf Plot	A plot which uses the digits of the data values to show the shape and distribution of the data set.	Gráfica de Tallo y Hoja	Un diagrama que utiliza los dígitos de los valores de datos para mostrar la forma y la distribución del conjunto de datos.
Straight Angle	An angle that measures 180°. 	Ángulo Recto	Un ángulo que mide 180°. 

Glossary ~ Glosario **155**

English	Definition	Spanish	Definición
Straight Edge	A ruler-like tool with no markings.	Borde Recto	Un gobernante como herramienta sin marcas.
Substitution Method	A method for solving a system of linear equations.	Método de Substitución	Un método para resolver un sistema de ecuaciones lineales.
Supplementary Angles	Two angles whose sum is 180°.	Ángulos Suplementarios	Dos ángulos cuya suma es 180°.
Surface Area	The sum of the areas of all the surfaces on a solid.	Área de la Superficie	La suma de las áreas de todas las superficies en un sólido.
System of Linear Equations	Two or more linear equations.	Sistema de Ecuaciones Lineales	Dos o más ecuaciones lineales.

T

English	Definition	Spanish	Definición
Term	A number or the product of a number and a variable in an algebraic expression; a number in a sequence.	Término	Un número o el producto de un número y una variable en una expresión algebraica; un número en una sucesión.
Terminating Decimal	A decimal that stops.	Decimal Terminado	Un decimal que para.
Theorem	A relationship in mathematics that has been proven.	Teorema	Una relación en las matemáticas que ha sido probada.
Theoretical Probability	The ratio of favorable outcomes to the number of possible outcomes.	Probabilidad Teórica	La proporción de resultados favorables a la cantidad de resultados posibles.
Third Quartile (Q3)	The median of the upper half of a data set.	Tercer Cuartil (Q3)	Mediana de la parte superior de un conjunto de datos.
Tick Marks	Equally divided spaces marked with a small line between every inch or centimeter on a ruler.	Marcas de Graduación	Espacios divididos igualmente marcados con una línea pequeña entre cada pulgada o centímetro en una regla.
Transformation	The movement of a figure on a graph so that it changes size or position.	Transformación	El movimiento de una figura en un gráfico de modo que cambia el tamaño o posición

Translation	A transformation in which a figure is shifted up, down, left or right. 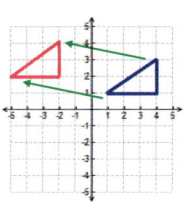	Traducción	Una transformación donde la figura se mudo arriba, abajo, a la izquierda o a la derecha. 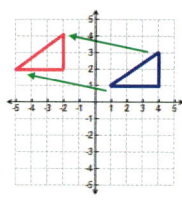
Transversal	A line that intersects two or more lines in the same plane.	Transversal	Una recta que interseca dos o más rectas en el mismo plano.
Trapezoid	A quadrilateral with exactly one pair of parallel sides. 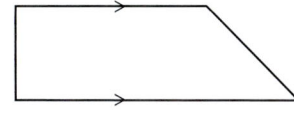	Trapezoide	Un cuadrilateral con exactamente un par de lados paralelos. 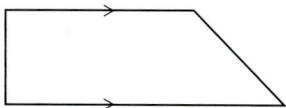
Tree Diagram	A display that organizes information to determine possible outcomes. 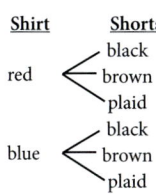	Diagrama de Árbol	Una pantalla que organiza la información para determinar los posibles resulatados. 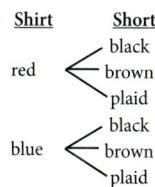
Trial	A single act of performing an experiment.	Prueba	Un solo intento de realizar un experimento.
Trinomial	An expression with three terms (i.e. $x^2 - 3x + 4$).	Trinomio	Una expreción que tiene tres terminos (es decir: $x^2 - 3x + 4$).
Two-Step Equation	An equation that has two different operations.	Ecuación de Dos Pasos	Una ecuación que tiene dos operaciones diferentes.
Two-Variable Data	A data set where two groups of numbers are looked at simultaneously.	Datos de dos Variables	Un conjunto de datos dónde dos grupos de números se observan simultáneamente.

| Two-Way Frequency Table | A table that shows how many times a value occurs for a pair of categorical data. | Tabla de Frecuencia Bidireccional | Una tabla que muestra cuántas veces aparece un valor de un par de datos categóricos. |

	Walk	
Dog Owners	Yes	No
Yes	15	20
No	25	20

	Paseo	
Perro Propietario	Si	No
Si	15	20
No	25	20

U-V-W

| Unit Rate | A rate with a denominator of 1. | Índice de Unidad | Un índice con un denominador de 1. |

| Univariate Data | Data that describes one variable (i.e., scores on a test). | Data Univariados | Datos que describen una variable (es decir: puntajes en una prueba). |

| Variable | A symbol that represents one or more numbers. | Variable | Un símbolo que representa uno o más números. |

| Vertex | The minimum or maximum point on a parabola. | Vértice | El mínimo o máximo punto en una parábola. |

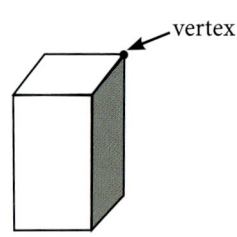 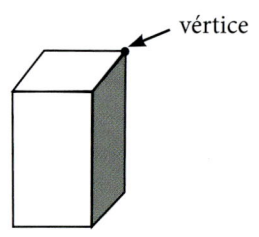

| Vertex of a Solid | The point where three or more edges meet. | Vértice de un Sólido | El punto donde tres o más bordes se encuentran. |

| Vertex of a Triangle | A point where two sides of a triangle meet. | Vértice de un Triángulo | Un punto donde dos lados de un triángulo se encuentran. |

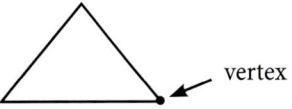 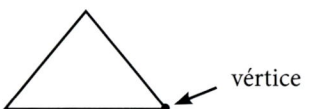

Vertex of an Angle	The common endpoint of the two rays that form an angle. 	Vértice de un Ángulo	El punto final en común de los dos rayos que forma un ángulo. 
Vertex Form	A quadratic function is in vertex form when written $f(x) = a(x - h)^2 + k$ where $a \neq 0$.	Forma De Vértice	Una función cuadrática es en forma general cuándo escrito $f(x) = a(x - h)^2 + k$ donde $a \neq 0$.
Vertical Angles	Non-adjacent angles with a common vertex formed by two intersecting lines. 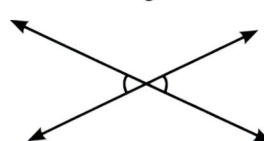	Ángulos Verticales	Ángulos no adyacentes con un vértice en común formado por dos rectas intersecantes. 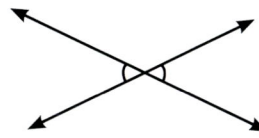
Vertical Line Test	A test used to determine if a graph represents a function by checking to see if a vertical line passes through no more than one point of the graph of a relation.	Examen Vertical De Línia	Un examen para determinar si una gráfica representa una función. Es utilizada para ver si una línia vertical que pasa a través de no más de un punto de la gráfica de una relación.
Volume	The number of cubic units needed to fill a three-dimensional figure.	Volumen	La cantidad de unidades cúbicas necesitadas para llenar un sólido.

X-Y-Z

x-Axis	The horizontal number line on a coordinate plane. 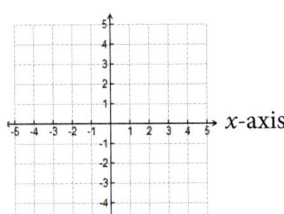	Eje-x, Eje de la x	La recta numérica horizontal en un plano de coordenadas. 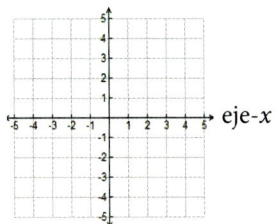

y-Axis	The vertical number line on a coordinate plane.	Eje-*y*, Eje de la *y*	La recta numérica vertical en un plano de coordenadas.

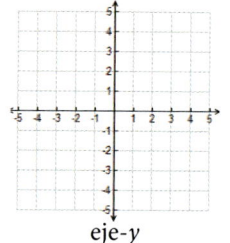

y-axis / eje-*y*

y-Intercept	The point where a graph intersects the *y*-axis.	Intersección *y*	El punto donde una gráfica interseca el eje-*y*.

Zero Pair	One positive integer chip paired with one negative integer chip.	Par Cero	Un chip entero positivo emparejado con un chip entero negativo.

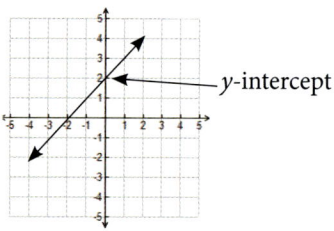

$$1 + (-1) = 0$$

Zero Product Property	If a product of two factors is equal to zero, then one or both of the factors must be zero.	Propiedad De Producto Cero	Si un producto de dos factores es iqual a cero, uno o ambos de los factores debe ser cero.
Zeros	The *x*-intercepts of a quadratic function.	Ceros	Las intersecciones-*x* de una función cuadratica.

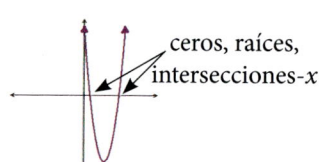

zeros, roots, *x*-intercepts / ceros, raíces, intersecciones-*x*

SELECTED ANSWERS

BLOCK 1

Lesson 1.1

1. ∠QUP or ∠PUQ or ∠U **3.** ∠BAT or ∠TAB or ∠A
5. See student work. **7.** See student work. **9.** See student work. **11.** 160° **13.** 113° **15.** 90° **17.** 50° **19.** See student work. **21.** See student work. **23.** ∠EAG or ∠GAE
25. ∠CAG or ∠GAC **27.** ∠BAE or ∠EAB **29.** always; See student work (∠PRM joins with ∠MRT to form ∠PRT, so ∠PRT must be greater than ∠PRM).

Lesson 1.2

1. ≈ 65°; acute **3.** ≈ 12°; acute **5.** 180°; straight
7. See student work. **9.** See student work. **11.** See student work. **13.** See student work. **15.** $x = 48$ **17.** $x = 50$
19. $x = 5$ **21.** $m\angle JAM = m\angle GEM = 85°$; See student work.
23. a) 17 **b)** Latoya is correct. See student work.
c) $17 < x < 39.5$ **25.** See student work. **27.** 69° **29.** 28°
31. See student work. **33.** See student work.

Lesson 1.3

1. complementary **3.** supplementary **5.** neither
7. neither **9.** $42 + x = 90$; $x = 48$ **11.** $74 + 2x = 180$; $x = 53$
13. $8x − 2 = 90$; $x = 11.5$ **15.** 123°; See student work.
17. 59°; See student work. **19.** $m\angle 1 = 90°$; $m\angle 2 = 90°$
21. $m\angle G = 37°$; $m\angle H = 143°$ **23.** $m\angle 1 = 65.6°$; $m\angle 2 = 24.4°$
25. always; See student work. **27.** See student work.
29. See student work. **31.** $m\angle JKL = 90°$, $m\angle XYZ = 90°$

Lesson 1.4

1. a) 50° **b)** 130° **c)** 130° **3.** See student work. **5.** See student work. **7.** See student work. **9.** vertical angles; $x = 22$
11. adjacent angles; $x = 17$ **13.** linear pair; $x = 40.5$
15. adjacent angles; $x = 19$ **17.** vertical angles; $x = 9$
19. ∠WXY and ∠TXY are supplementary so their sum is 180° not 90°. $m\angle TXY = 137°$ **21.** Vertical angles are formed by intersecting lines; they do not share a side so they are not adjacent. **23.** right angle **25.** supplementary angles
27. congruent angles **29.** linear pair and supplementary angles

Lesson 1.5

1. never; See student work. **3.** always; See student work.
5. sometimes; See student work. **7.** Yes, these lengths form a triangle. **9.** Yes, these lengths form a triangle. **11.** No, these lengths do not form a triangle. **13.** No, only one pair of corresponding angles are congruent. **15.** See student work. Yes, the triangles are congruent. **17.** between 5 *in* and 29 *in*
19. between 9.5 *ft* and 74.5 *ft* **21.** No, corresponding angles and sides are not congruent. **23.** See student work.
25. See student work. **27.** They could measure the bottom two angles of every sail to verify the corresponding angles are congruent. If they are, then the sails are similar by the Angle-Angle Similarity Theorem. **29.** ∠K or ∠5 **31.** See student work. **33.** See student work.

Block 1 Review

1. 51° **3.** 39° **5.** ∠XYZ, ∠ZYX, ∠Y or ∠5 **7.** See student work. **9.** acute **11.** right **13.** See student work.
15. $35 + 2x = 31$; $x = −2$ **17. a)** 76 **b)** x is between −14 and 76
c) No, x must be between 76 and 166. See student work.
19. neither **21.** $x = 41$ **23.** $x = 38$ **25.** See student work. $x = 53$ **27.** $a = 135°$, $b = 45°$, $c = 135°$ **29.** vertical angles; $x = 133$ **31.** vertical angles; $x = 20$ **33.** linear pair or supplementary angles; $x = 16.5$ **35.** No, the lengths cannot form a triangle. **37.** Yes, the lengths can form a triangle.
39. between 4.5 *cm* and 9.1 *cm* **41.** Yes, the triangles are similar by the Angle-Angle Similarity Theorem. **43.** false; If two angles in one triangle are congruent to two angles in another triangle, then the triangles are similar to one another.

BLOCK 2

Lesson 2.1

1. 30 cm^2 **3.** 55 in^2 **5.** 17.5 in^2 **7.** See student work. 20 cm^2
9. See student work (36 in^2). **11.** $x = 10$ cm **13.** $x = 9$ meters
15. $x = 5$ meters **17.** The area is 20 cm^2 so it needs to be substituted on the left side of the equation. The width is 8 *cm*.
19. a) 32 ft^2 **b)** 4 ft^2 and 16 ft^2 **c)** One flower bed gives more area; See student work. **21.** 640 square inches **23.** See student work (16 square units). **25.** See student work (20 square units). **27.** Answers may vary; between 18 units and 19 units is a close estimate. The exact perimeter cannot be found because the length of the diagonal side can only be estimated. **29.** 70° **31.** 30° **33.** See student work. **35.** $x = 23$
37. $x = 8$

Selected Answers **161**

Lesson 2.2

1. 20 in^2 **3.** 45 cm^2 **5.** 15.45 yd^2 **7.** 162 in^2 **9.** 70 square feet
11. 200 square inches **13.** $h = 10$ ft **15.** $b_1 = 8.2$ meters
17. $b_2 = 8$ m **19.** Answers may vary (must be two numbers whose sum is 20). **21. a)** 42 m^2 **b)** 66.5 m^2 **c)** 108.5 m^2
d) 108.5 m^2. Yes, the answers are both 108.5 m^2. See student work. **23. a)** red → 51 ft^2, green → 51 ft^2 **b)** 102 ft^2
c) parallelogram **d)** 102 ft^2 **e)** Yes; answers may vary (e.g. both methods find the area of the same shape). One method uses one shape, a parallelogram, and the other method adds two trapezoids together to form the parallelogram.
25. 24.5 m^2 **27.** $x = 6$ in **29.** $x = 5.5$ m

Lesson 2.3

1. Answers may vary (e.g., trace around a jar lid, wheel, coin...) **3.** See student work. **5.** See student work. **7.** See student work. **9.** $\overline{BD}$, $\overline{AB}$ or $\overline{BC}$ **11.** ∠ABC, ∠CBD or ∠ABD
13. three of the following: $\overline{VC}$, $\overline{VI}$, $\overline{VE}$, $\overline{VS}$ or $\overline{VR}$ **15.** two of the following: $\overline{IE}$, $\overline{RS}$ or $\overline{SL}$ **17.** 56 m **19.** $9\frac{1}{3}$ inches
21. 7.5 yd **23.** $x = 60$ **25.** $x = 127$ **27.** $x = 20$ **29.** 150°;
See student work. **31.** use the central angle sum to show $x + 100 + 80 + 100 = 360$ so $x = 80$ **33.** 120 cm^2
35. 30 square units **37.** 30 ft^2

Lesson 2.4

1. 31.4 yards **3.** 21.98 ft **5.** 8.164 m **7. a)** ≈ 53.38 inches
b) About 53.38 inches because one rotation is the same as the circumference of the tire. **9.** ≈ 7,850 feet **11.** $d = 4$ ft
13. $d = 68$ in **15.** $r = 10$ miles **17.** 1 mile **19.** 54 inches;
See student work. **21.** 3.14 is an approximation of pi
23. See student work. **25.** 6 meters **27.** 9 inches

Lesson 2.5

1. 50.24 ft^2 **3.** 314 mm^2 **5.** Theo multiplied 16 by 2 instead of squaring 16. The area of the pizza is about 803.84 in^2.
7. 1,385 square miles **9.** See student work. The two small windows have an area of about 6.28 ft^2 and the one large window has an area of about 12.56 ft^2 so she should use the one large window with a radius of 2 feet. **11.** 81π in^2
13. 64π mm^2 **15.** The circles have equal areas because they both have a radius of 2.5 inches. See student work.
17. a) 18 inches **b)** 254.34 square inches **19.** ≈ 4.91 m^2;
See student work. **21. a)** 8π cm^2 **b)** 4π cm^2 **c)** 2π cm^2
d) $\frac{16}{3}$π cm^2 or $5\frac{1}{3}$π cm^2 **23.** two of the following: $\overline{HC}$, $\overline{HI}$ or $\overline{HP}$ **25.** $\overline{ZU}$ **27.** ⊙H

Lesson 2.6

1. 3.14; this will be close enough for a good estimate.
3. $\frac{22}{7}$; 42 is a multiple of 7 **5.** C ≈ 25.13 cm; A ≈ 50.27 cm^2
7. C ≈ 39.27 cm; A ≈ 122.72 cm^2 **9.** C ≈ 88 ft; A ≈ 616 ft^2
11. C ≈ 44 yd; A ≈ 154 yd^2 **13.** C ≈ 66 mi; A ≈ 346.5 mi^2
15. 120 feet **17.** 113.04 ft^2 **19.** 49π cm^2; See student work.
21. ≈ 72.26 ft **23.** A = $\frac{1}{2}bh$; triangle **25.** A = $\frac{1}{2}h(b_1 + b_2)$; trapezoid **27.** A = s^2; square

Lesson 2.7

1. ◯ − ◯ **3.** ▢ + ◗ **5.** ◯ − △ − ▽
7. 91.5 square units **9.** 62.8 cm^2 **11.** 7.23 m^2 **13.** 239.5 m^2
15. Keisha used a full circle instead of a half circle when finding the total area. The area is about 89.12 cm^2.
17. 47.98 ft **19.** 21.86 square inches **21.** 10 rectangles; See student work. **23.** ≈ 10.09 in^2 **25.** $x = 65$ **27.** See student work (i.e., 24.5 square units).

Lesson 2.8

1. 5 : 1 **3.** 7 : 3 **5.** 2 : 11 **7.** $\frac{2}{5}$ **9.** 1 : 16 **11.** 16 : 9
13. $d = 12$ inches **15.** 20 square meters **17.** 35.5 cm
19. All circles are the same shape but may be different sizes.
21. 56.5 cm^2 **23.** 140.14 cm^2 **25.** 50.24 cm^2

Lesson 2.9

1. 10 cm^2 **3.** ≈ 14.47 ft^2 **5.** ≈ 2.75 in^2 **7. a)** 270° **b)** 37.68 in^2;
See student work. **9.** 452.16 m^2 **11.** ≈ 21.20 in^2
13. ≈ 0.37 cm^2 **15.** 3π yd^2 **17.** ≈ 41.87 in^2 **19.** ≈ 2.09 in^2
21. 5 units

Block 2 Review

1. 55 ft^2 **3.** 49 yd^2 **5.** 36 ft^2 **7.** $x = 6$ m **9.** 60 square units
11. 63 cm^2 **13.** $b_1 = 10$ in **15.** $3\frac{1}{3}$ in **17.** two of the following: $\overline{HN}$, $\overline{HO}$, $\overline{HS}$ or $\overline{HE}$ **19.** two of the following: $\overline{KI}$, $\overline{CR}$ or $\overline{NS}$
21. $\overline{NS}$ **23.** $x = 106$ **25.** See student work. **27.** 75.36 ft
29. 785 ft **31.** 8 in **33.** $r = 4$ yd **35.** 28.26 cm^2 **37.** 132.665 m^2
39. 5,024 ft^2 **41.** 4π m^2 **43.** Pierre used the diameter in the area formula instead of the radius. The area is 16π ft^2.
45. $\frac{22}{7}$; the diameter is a multiple of 7; C ≈ 22 in **47.** $\frac{22}{7}$; radius is a multiple of 7; A = 154 m^2 **49. a)** Calculator π because it uses more digits after the decimal point for a more accurate estimate. **b)** 69.40 ft^2 **51.** Answers may vary. **53.** 64.8 ft^2
55. 512.52 cm^2 **57.** 120 in^2 **59.** 2 : 5 **61.** 4 : 1 **63.** 16 : 1
65. $x = 10.98$ cm **67.** 98 in^2 **69.** 96.79 in^2 **71.** 14.13 in^2
73. 50.24 cm^2

BLOCK 3

Lesson 3.1

1. rectangular prism **3.** square or rectangular pyramid
5. cylinder **7.** cone **9.** triangular prism **11.** true
13. false; The faces of a pyramid are triangles. **15.** true
17. false; Only hexagonal prisms have 12 vertices. **19.** false; pyramids are named by their bases **21.** lateral faces: 5; bases: 1; edges: 10; vertices: 6; **23. a)** rectangular prism
b) 6 **c)** 8 **d)** 12 **e)** square pyramid **f)** 5 **g)** 5 **h)** 8 **i)** cylinder
j) 2 **k)** 0 **l)** 0 **m)** octagonal prism **n)** 10 **o)** 16 **p)** 24
25. total faces = lateral faces + base(s) **27.** cylinder
29. triangular pyramid **31.** pentagonal pyramid
33. Answers may vary (e.g., cube and heptagonal pyramid, pentagonal prism and octagonal prism).
35. C ≈ 88 cm; A ≈ 616 cm^2 **37.** C ≈ 52.81 m; A ≈ 222.09 m^2

Lesson 3.2

1.-17. See student work. **19.** Piper counted each vertex and edge in the net, forgetting when the net is folded into a pyramid, edges and vertices will be shared by faces. There are 5 vertices, 8 edges, 4 lateral faces and 1 base.
21. vertices: 6; edges: 9; lateral faces: 3; bases: 2
23. vertices: 0; edges: 0; lateral faces: 0; bases: 2
25. 48 square units

Lesson 3.3

1. square **3.** circle **5.** triangle **7.** rectangle **9.** rectangle
11. a) square **b)** rectangular prisms **13. a)** 96 in^2; See student work. **b)** ≈ 50.24 in^2; See student work. **15.** pentagon, congruent **17.** hexagon, similar **19.** cone or any pyramid
21. triangle **23.** oval **25.** See student work. **27.** See student work. **29.** $h = 12$ in

Lesson 3.4

1. a) See student work. **b)** 15 m^2, 15 m^2, 24 m^2, 24 m^2, 40 m^2, 40 m^2 **c)** 158 m^2 **3.** 82 ft^2 **5.** 987 u^2 **7.** 1,451.6 cm^2
9. Zuleyma only included one base instead of two bases in her work. The surface area is 460 in^2. **11.** 435 ft^2 **13.** 6 ft^2; See student work. **15. a)** 416 in^2 **b)** There is overlap when wrapping a gift. **17.** $x = 15$ feet; See student work. **19.** 30 ft^2
21. 45 cm^2 **23.** 462 in^2

Lesson 3.5

1. 160 in^3 **3.** 147 cm^3 **5. a)** 135 ft^3 **b)** 67.5 ft^3 **c)** The rectangular prism has twice the volume as the triangular prism; See student work. **7. a)** 210 in^3 **b)** 210 pieces **c)** 1,680 pieces; See student work. **9.** 230 in^3 **11.** 72 ft^3
13. 39 in^3 **15. a)** 4.5 ft^3 **b)** 7,776 in^3 **c)** Answers may vary.
17. Answers may vary. The product of the three dimensions must equal 36 in^3. **19.** 44 in **21.** 440 mm **23.** 1,406.72 in^2

Lesson 3.6

1. 8 **3.** 6 **5.** 3 **7. a)** See student work. **b)** 16 cm^2, 20 cm^2, 20 cm^2, 20 cm^2, 20 cm^2 **c)** 96 cm^2 **9.** 270 ft^2 **11.** 115.5 u^2
13. 216 in^2 **15.** 147.7 in^2 **17.** 979.61 ft^2 **19.** 176 cm^2
21. 42 square feet; See student work. **23.** 184 in^2
25. 12 edges **27.** 8 lateral faces

Lesson 3.7

1. The volume of a pyramid is one-third the volume of a prism OR the prism is three times the volume of the pyramid. **3.** 360 cm^3 **5.** 30 minutes **7.** 88 in^3 **9.** $133\frac{1}{3}$ ft^3
11. 281.216 m^3 **13.** 400 cm^3 **15.** 2.5 in^3 **17.** 54 m^2
19. circle; similar **21.** rectangle; congruent

Block 3 Review

1. 6 faces, 4 lateral faces, 2 bases, 12 edges, 8 vertices
3. 4 faces, 3 lateral faces, 1 base, 6 edges, 4 vertices **5.** sphere
7. triangular pyramid **9.** Answers may vary (cylinder or prism). **11.** never; See student work. **13.** See student work.
15. See student work. **17.** See student work. **19.** See student work. **21.**

Solid	What shape is formed by a slice parallel to the base?	Is the parallel slice similar or congruent to the base?	What shape is formed by a perpendicular slice to the base?
Cylinder	a. Circle	b. Congruent	c. Rectangle
Cone	d. Circle	e. Similar	f. Triangle
Triangular Prism	g. Triangle	h. Congruent	i. Rectangle
Octagonal Pyramid	j. Octagon	k. Similar	l. Triangle

23. triangular pyramid; See student work. **25.** 190 in^2
27. 766 cm^2 **29.** 276 u^2 **31.** 2.5 in^2; See student work.
33. 160 ft^3 **35.** $78\frac{1}{8}$ in^3 **37.** the triangular container; (rectangular container = 120 in^3, triangular container = 156 in^3); See student work. **39.** Answers may vary. **41.** 4 in^2
43. 919,880 ft^2 **45.** The pyramid is one-third the volume of the prism or the prism holds three times the volume of the pyramid. **47.** See student work. **49.** 72 in^3 **51.** 48 cm^3

INDEX

A

Angle, 3
  acute, 8
  adjacent, 3, 20
  classifying, 8
    Explore! Classify an Angle, 8
  complementary, 14
    Explore! Complementary vs.
        Supplementary, 13
  congruent, 9
    notation, 9
  measuring of, 4
  naming, 3
  obtuse, 8
  right, 8
    notation, 9
  straight, 9
  supplementary, 14
    Explore! Complementary vs.
        Supplementary, 13
  vertical, 19, 20

Angle-Angle Similarity Rule, 25
  Explore! Knowing Three Measures:
    Part 1, Three Angles, 24

Area, 37
  circle, 57, 65
    Explore! Circle Areas, 56
  finding a missing measure, 38
  parallelogram, 37, 65
  sector, 74
  trapezoid, 42, 65
    Explore! A Fromula for a Trapezoid, 42
  triangle, 37, 65
  rectangle, 37, 65
  square, 37, 65

B

Base, 87
  of a solid, 87
  of a trapezoid, 42

C

Career Focus
  Mason, 124
  Nursery Manager, 34
  Safety and Health Professional, 84

Center, 47

Central angle, 47

Central angle sum, 48

Chord, 47

Circle, 47
  area, 65
  circumference, 52-53
    Explore! A Special Ratio, 52
  similarity, 70
    Explore! Stepping Stones, 69

Circumference, 52, 53

Complementary angles, 14
  Explore! Complementary vs.
    Supplementary, 13

Composite figures, 65

Cone, 87,
  slicing a cone, 97

Cylinder, 87
  slicing a cylinder, 97

D

Degree, 4

Diameter, 47, 48

E

Edge, 88

Explore!
  A Special Ratio, 52
  Circle Areas, 56
  Complementary vs. Supplementary, 13
  Cutting Clay, 96
  Knowing Three Measures, Part 1: Three
    Angles, 24
  Knowing Three Measures, Part 2: Three
    Sides, 24
  Netting a Solid, 92
  Pyramid vs. Prism, 115
  Take Your Pick, 101
  Tent Making, 111

F

Face, 87

G

H

Height
  pyramid, 87

I

J

K

L

Lateral face, 88, 101

Lateral area, 102
  prism, 102
  pyramid, 112

Lateral surface area, *see* Lateral area

Linear pair, 19, 20

M

N

Net, 92
  Explore! Netting a Solid, 92

O

Obtuse angle, 8

P

Perimeter, 38

Pi, 53
  Explore! Which Pi?, 61]
  choosing common estimates, 62

Polygon, 87

Prism, 87
  drawing a prism, 93
  naming a prism, 88
  slicing a prism, 97

Proportion, 70

Protractor, 4

Pyramid, 87
  drawing a pyramid, 93
  naming a pyramid, 88
  slicing a pyramid, 97

Q

R

Radius, 47

Ratio, 69

Ray, 3

Right angle, 8

S

Sector, 74
   area of, 74

Side Length Inequality Rule, 26
   Explore! Knowing Three Measures,
      Part 2: Three Sides, 24

Side-side-side triangle congruence, 27

Slant height, 111
   pyramid, 111, 115

Slice, 97
   Explore! Cutting Clay, 96

Sphere, 87
   slicing a sphere, 97

Straight angle, 9

Similar Triangles, 58

Supplementary Angles, 14
   Explore! Complementary vs.
      Supplementary, 13

Surface area, 101
   prism, 102
      Explore! Take Your Pick, 101
   regular pyramid, 112
      Explore! Tent Making, 111

T

Total surface area, *see* surface area

Trapezoid, 42
   area of, 42

U

V

Vertex, 3, 88

Vertical Angles, 19, 20
   Explore! Circle Areas, 56

Volume, 106
   prism, 107
      Explore! Cutting Corners, 106
   pyramid, 116
      Explore! Pyramid vs. Prism, 115

W X Y Z

PROBLEM-SOLVING

UNDERSTAND THE SITUATION

- Read then re-read the problem.
- Identify what the problem is asking you to find.
- Locate the key information.

PLAN YOUR APPROACH

Choose a strategy to solve the problem:

- Guess, check and revise
- Use an equation
- Use a formula
- Draw a picture
- Draw a graph
- Make a table
- Make a chart
- Make a list
- Look for patterns
- Compute or simplify

SOLVE THE PROBLEM

- Use your strategy to solve the problem.
- Show all work.

ANSWER THE QUESTION

- State your answer in a complete sentence.
- Include the appropriate units.

STOP AND THINK

- Did you answer the question that was asked?
- Does your answer make sense?
- Does your answer have the correct units?
- Look back over your work and correct any mistakes.

DEFEND YOUR ANSWER

Show that your answer is correct by doing one of the following:

- Use a second strategy to get the same answer.
- Verify that your first calculations are accurate by repeating your process.

SYMBOLS

Algebra and Number Operations

SYMBOL	MEANING
+	Plus or positive
−	Minus or negative
$5 \times n, 5 \cdot n, 5n, 5(n)$	Times (multiplication)
$3 \div 4, 4\overline{)3}, \frac{3}{4}$	Divided by (division)
=	Is equal to
≈	Is approximately
<	Is less than
>	Is greater than
%	Percent
$a : b$ or $\frac{a}{b}$	Ratio of a to b
$5.\overline{2}$	Repeating decimal (5.222…)
≥	Is greater than or equal to
≤	Is less than or equal to
x^n	The nth power of x
(a, b)	Ordered pair where a is the x-coordinate and b is the y-coordinate
±	Plus or minus
$\sqrt{x}$	Square root of x
≠	Not equal to
$x \stackrel{?}{=} y$	Is x equal to y?
$|x|$	Absolute value of x
P(A)	Probability of event A

Geometry and Measurement

SYMBOL	MEANING
≅	Is congruent to
~	Is similar to
∠	Angle
$m\angle$	Measure of angle
△ABC	Triangle ABC
$\overline{AB}$	Line segment AB
$\overrightarrow{AB}$	Ray AB
AB	Length of AB
π	Pi (approximately $\frac{22}{7}$ or 3.14)
°	Degree